Telegeoprocessing

Series in Remote Sensing

ISSN: 2010-2933

Published:

Vol. 5 *Telegeoprocessing*
by Yong Xue, Xiran Zhou and Sheng Zhang

Vol. 4 *Wildland Fire Danger Estimation and Mapping: The Role of Remote Sensing Data*
edited by Emilio Chuvieco

Vol. 3 *Analysis of Multi-Temporal Remote Sensing Images: Proceedings of the Second International Workshop on the Multitemp 2003*
edited by Paul Smits and Lorenzo Bruzzone

Vol. 2 *Analysis of Multi-Temporal Remote Sensing Images: Proceedings of the First International Workshop on Multitemp 2001*
edited by Lorenzo Bruzzone and Paul Smits

Vol. 1 *Soft Computing in Remote Sensing Data Analysis: Proceedings of the International Workshop*
edited by Elisabetta Binaghi, Pietro Alessandro Brivio and Anna Rampini

Series in Remote Sensing Vol. 5

Telegeoprocessing

Yong Xue, Xiran Zhou & Sheng Zhang

China University of Mining and Technology, China

NEW JERSEY • LONDON • SINGAPORE • BEIJING • SHANGHAI • HONG KONG • TAIPEI • CHENNAI • TOKYO

Published by

World Scientific Publishing Co. Pte. Ltd.
5 Toh Tuck Link, Singapore 596224
USA office: 27 Warren Street, Suite 401-402, Hackensack, NJ 07601
UK office: 57 Shelton Street, Covent Garden, London WC2H 9HE

Library of Congress Cataloging-in-Publication Data
Names: Xue, Yong (Geoscientist), author. | Zhou, Xiran, author. | Zhang, Sheng, 1971– author.
Title: Telegeoprocessing / Yong Xue, Xiran Zhou & Sheng Zhang,
China University of Mining and Technology, China.
Description: Singapore : World Scientific, 2023. | Series: Series in remote sensing,
2010-2933 ; Vol. 5 | Includes bibliographical references and index.
Identifiers: LCCN 2022025787 | ISBN 9789811262173 (hardcover) |
ISBN 9789811262180 (ebook) | ISBN 9789811262197 (ebook other)
Subjects: LCSH: Geographic information systems. | Global Positioning System. |
Geospatial data. | Telecommunications systems.
Classification: LCC G70.212 .X84 2023 | DDC 910.285--dc23/eng20221128
LC record available at https://lccn.loc.gov/2022025787

British Library Cataloguing-in-Publication Data
A catalogue record for this book is available from the British Library.

For any available supplementary material, please visit
https://www.worldscienti ic.com/worldscibooks/10.1142/13019#t=suppl

Desk Editors: Balasubramanian Shanmugam/Steven Patt

Typeset by Stallion Press
Email: enquiries@stallionpress.com

Printed in Singapore

Preface

This is a cross-disciplinary book aimed at teachers and researchers in both universities and research organizations; indeed, it is for anyone who is interested in the impact of Earth observation, Big Data, and Geoinformatics in civil communities and human societies. Our aim is to stimulate debate so that we can make better use of new technologies in contemporary society. What we are proposing in this book is not a new application of remotely sensed data or a new way of handling data for a particular problem or application rather it is an attempt to suggest a synthesis of remote sensing with a number of other rapidly developing technologies. It includes questions related to the access to the data, to the integration of remotely sensed data with data from other sources, to the analysis and interpretation of the data to extract geophysical or environmental information, and to the distribution of information.

Telegeoprocessing can be defined as a new discipline based on real-time spatial databases updated regularly by means of telecommunications systems in order to support problem solving and decision-making at any time and any place. Telegeoprocessing is the integration of remote sensing, Geographic Information System (GIS), Global Navigation Satellite System (GNSS), Big Data, and Telecommunication. A discussion of Telegeoprocessing is timely. This book intends to bring together most of the key issues involved in research in novel systems in Telegeoprocessing. There has been a startling growth in the use of Telegeoprocessing.

Telegeoprocessing can be categorized as WebGIS (an integration of Internet and GIS), Mobile Geoprocessing (Wireless Mapping — an integration of mobile computing and GIS), and TeleGIS (an integration of

GIS and other telecommunication techniques). This book presents a comprehensive introduction to the problems encountered in Telegeo-processing engineering and the major technologies and standards related to designing an integrated, fully functional Telegeoprocessing system based on the latest multimedia and telecommunication technologies. Overall attention will be given to systems based on remote sensing and GNSS and all other technologies for sending data from sensors to GIS systems by means of multimedia telecommunications.

Telegeoprocessing can be used in many different contexts and with many different intentions. It has become a major demand for countries to seize the strategic commanding heights, safeguard national security, protect maritime rights and interests, effectively implcmcnt emergency rescue of major disasters, and provide all-round commercial public services.

Y. Xue

About the Author

Professor Dr. Yong Xue is an expert in quantitative remote sensing and Big Earth data, an Academician of the International Academy for Europe and Asia, a Corresponding Academician of the International Academy of Astronautics, and a Chartered Physicist. He received his BSc degree in Space Physics from the Department of Geophysics, Peking University, in 1986, his MSc degree from the Institute of Remote Sensing and Geoinformation, Peking University, in 1989, and his Doctor of Philosophy (PhD) from the Department of Applied Physics, University of Dundee, in 1995.

He is currently a Chair Professor in the School of Environment and Spatial Informatics, China University of Mining and Technology (CUMT), the team leader of the Division of Photogrammetry and Remote Sensing at CUMT, the deputy chair of "Chinese National Committee of the International Society for Digital Earth — Digital Energy Commission", the deputy director of Observation and Research Station of Jiangsu Jiawang Resource Exhausted Mining Area Land Restoration and Ecological Succession, Ministry of Education, China. He is also the director of the Collaborative Cross-Disciplines Research Center of Telegeoprocessing at CUMT, the director of the Remote Sensing Big Data Center for Energy and Environment at CUMT, and the director of the Center for Intelligent Remote Sensing Analysis, Institute of Artificial

Intelligence at CUMT. He has published more than 310 peer-reviewed papers (more than 260 corresponding and first-author papers). He has published two academic monographs (co-author) and more than 320 invited conference presentations. He is currently on the editorial board of the Journal *International Journal of Remote Sensing* and the editor of the World Scientific Publishing Remote Sensing book series (2001–).

Professor Dr. Yong Xue's contact details are listed below.

School of Environment and Spatial Informatics,
China University of Mining and Technology,
No. 1 University Road, Xuzhou 221116,
Jiangsu Province, PR China

Email: yx9@hotmail.com

Contents

Chapter 1

Remote Sensing and Real-Time Processing in Remote Sensing

1.1 Remote Sensing and Remote Sensing Process

1.1.1 *Remote sensing*

Remote sensing is the science of obtaining and interpreting information about the Earth from distant vantage points using sensors, usually by/from satellites or aircraft. Sensors mounted on these platforms capture detailed pictures of the Earth that reveal features not apparent to the naked eye. The science of remote sensing in its broadest sense includes aerial, satellite, and spacecraft observations of the surfaces and atmospheres of the planets in our solar system, though the Earth is obviously the most frequent target of study. There are two kinds of remote sensing. Passive sensors detect natural energy (radiation) that is emitted or reflected by the object or scene being observed. Reflected sunlight is the most common source of radiation measured by passive sensors. Active sensors, on the other hand, provide their own source of energy to illuminate the objects they observe. An active sensor emits radiation in the direction of the target to be investigated. The sensor then detects and measures the radiation that is reflected or backscattered from the target.

The four most important considerations in remote sensing measurement are as follows: Reliability of observations which is often the most critical assessment that needs to be considered in any research. Validity addresses the remote sensing measurement issues regarding the nature, meaning, or definition of a concept or variable. Often multiple interpretations are made, which necessitate operational definitions. Precision refers to the level of exactness associated with remote sensing measurement and is often determined by the calibration of the remote sensing sensors. Accuracy refers to the extent of system-wide bias in the remote sensing imaging process so that distorted measures of reality are not obtained. It is possible to have precise yet inaccurate measurements. When evaluating the quality of data or the selection of a measurement scale, five basic sources of error can arise which are as follows: calculation error, measurement error, specification error, sampling error, and random noise.

Today, there are seemingly countless benefits and applications of space technology. Satellites, for instance, are becoming critical for everything from internet connectivity and precision agriculture to border security and archaeological study. The instruments aboard orbiting satellites transmit data such as temperature, electromagnetic spectrum, energy output, heat, and light from the Earth's surface and atmosphere. Once captured, the images are passed on to analysts who interpret the data, extract information, and use it to answer questions. Satellite data, in conjunction with analysis of Earth system data from other sources, provide scientific teams with enough information to predict many of the Earth's processes. This information may be used to map forests, detect pollution, measure elevation, locate a diseased crop, and answer a variety of questions.

Thousands of satellites orbit the Earth. Right now, there are nearly 6,000 satellites circling our tiny planet. About 60% of those are defunct satellites — space junk — and roughly 40% are operational. Earth observation (EO) satellites are satellites specifically designed to observe Earth from orbit. Over the coming decade, it's estimated by Euroconsult that 990 satellites will be launched every year. This means that by 2028, there could be 15,000 satellites in orbit. Second to communications, 27% of commercial satellites have been launched for EO purposes, including environmental monitoring and border security. Operational Remote Sensing Satellites are listed as following:

Commercial high-spatial resolution

- SuperView-1 (http://www.carsa.org.cn/upload/5a6ef971956fd.pdf)
- Jilin-1 (http://www.jl1.cn/)
- WorldView (http://worldview3.digitalglobe.com/)
- Disaster Monitoring Constellation (https://earth.esa.int/eogateway/missions/dmc)

- IKONOS and OrbView-2 (https://www.eoportal.org/satellite-missions/orbview-1#orbview-1-formerly-microlab-1)
- QuickBird (https://www.eoportal.org/satellite-missions/quickbird-2)
- SPOT (https://spot.cnes.fr/en/SPOT/index.htm)
- EROS (https://www.imagesatintl.com/)
- RapidEye (https://earth.esa.int/eogateway/missions/rapideye)
- FORMOSAT-2 (https://www.nspo.narl.org.tw/inprogress.php?c=20022501&ln=en)

Earth Observing System

- Aqua (EOS PM-1) (http://aqua.nasa.gov/)
- Aura (http://aura.gsfc.nasa.gov/)
- GRACE (http://www.csr.utexas.edu/grace/)
- Jason 1 (https://www.nasa.gov/centers/jpl/missions/jason.html)
- Terra (EOS AM-1) (http://terra.nasa.gov/)
- ALOS (https://global.jaxa.jp/projects/sat/alos4/)

Geostationary Operational Environmental Satellite (GOES)

- GOES 12 (http://www.goes.noaa.gov/)
- GOES-N (https://www.nasa.gov/mission_pages/goes-n/index.html)

Geosynchronous Meteorological Satellite (GMS)/Himawari

- GMS/Himawari (https://global.jaxa.jp/projects/sat/gms/index.html)
- MTSAT-1R (https://www.jma.go.jp/jma/jma-eng/satellite/introduction/MTSAT_series.htm)

Landsat program

- Landsat 7
- Landsat 8
- Landsat 9 (http://landsat.gsfc.nasa.gov/)

Argentina Space Agency (CONAE)

- SAC-C (https://www.argentina.gob.ar/ciencia/conae/misiones-satelitales/sac-c)
- SAOCOM (https://www.argentina.gob.ar/ciencia/conae/misiones-espaciales/saocom)

Centre National d'Études Spatiales (CNES)

- SPOT (https://spot.cnes.fr/en/SPOT/index.htm)

ESA

- Envisat (https://earth.esa.int/eogateway/missions/envisat)
- ERS (https://earth.esa.int/eogateway/missions/ers)
- GeoEye-1 (https://earth.esa.int/eogateway/missions/geoeye-1)
- Pleiades (https://earth.esa.int/eogateway/missions/pleiades)
- Sentinel (https://sentinel.esa.int/web/sentinel/home)
- Meteosat (http://www.eumetsat.int)

ISRO

- IRS P6 (Resourcesat 1) (http://www.isro.gov.in/pslvc5/index.html)
- IRS P5 (Cartosat 1) (http://www.isro.org/Cartosat/Page3.htm)
- IRS P4 (Oceansat 1) (http://www.isro.org/irsp4.htm)

NASA

- Thermosphere Ionosphere Mesosphere Energetics and Dynamics (TIMED) (http://www.timed.jhuapl.edu/WWW/index.php)
- TOPEX/Poseidon (https://eospso.nasa.gov/missions/topexposeidon)
- Upper Atmosphere Research Satellite (UARS) (https://www.nasa.gov/mission_pages/uars/index.html)

NOAA

- NOAA-N (https://www.nasa.gov/mission_pages/noaa-n/main/)

Swedish National Space Board

- Munin (http://munin.irf.se/)

RADARSAT series

- RADARSAT-2 (https://www.asc-csa.gc.ca/eng/satellites/radarsat/)

Feng Yun (FY) series

- FY GEO series (http://fy4.nsmc.org.cn/nsmc/cn/satellite/FY4.html)
- FY LEO series (http://www.nsmc.org.cn/nsmc/en/satellite/FY3.html)

Gaofen (GF) High-resolution Imaging Satellite/CHEOS series

- GF-1, -2, -3, -4, -5, -6, -7 (http://www.sastind.gov.cn/n25770/index.html)

The small satellite renaissance began in the 1980s and is changing the economics of space. Technological trends have supported the advancement of small satellites in the 1–500 kg range. The number of countries actively participating has grown substantially during the past years. Satellite constellations (groups of satellites working in concert) are emerging as a powerful and effective application.

Future small satellite systems for both EO as well as deep-space exploration are greatly enabled by the technological advances in deep submicron microelectronics technologies. Whereas these technological advances are being fueled by the commercial (non-space) industries, more recently there has been an exciting new synergism evolving between the two otherwise disjoint markets. In other words, both the commercial and space industries are enabled by advances in low-power, highly integrated, miniaturized (low-volume), lightweight, and reliable real-time embedded systems.

To achieve an effective operational revisit frequency, it is necessary to have several satellites coordinated in a constellation and to integrate different small satellite constellations. Historically, this desirable capability has been too expensive to implement because of the cost of the satellites. However, small low-cost satellites are changing the equation. With a cost one-tenth of the original satellites, and a much more rapid build–launch cycle, these are bringing EO into a fully operational phase. The small satellites cannot carry as many instruments as the giant research satellites but they can put multiples of the same instrument into a coordinated orbit at an affordable cost. This enables the provision of a daily revisit service that reduces the impact of cloud and enables business to build new services using EO data. Examples of satellite constellations include the Global Positioning System (GPS), Galileo and GLONASS constellations for navigation and geodesy, RASAT, the Disaster Monitoring Constellation (DMC), RapidEye, COSMO-SkyMed (COnstellation of small Satellites for the Mediterranean basin Observation), and Huanjing constellation (the Small Satellite Constellation for Environment Protection and Disaster Monitoring) for remote sensing.

Small satellite constellations open a new era in sustainable EO. The next generation of small satellite constellation will have enhanced sensors, providing continuity data and extraordinary pace. The next improvement in the DMC is the increase in sensor resolution from 32 m ground sample distance (GSD) to 2 m GSD. The GSD refers to the size of the pixels in a digital orthophoto, expressed in ground units. For cxamplc, if an orthophoto has a 1.0 m GSD, each pixel represents a ground area measuring 1 m × 1 m. The combination of multiple spacecraft in constellation provides for daily coverage at a resolution that enables effective monitoring of the rapidly changing environment.

Although most remote sensors collect their data using the basic principles described above, the format and quality of the resultant data vary widely. These variations are dependent upon the resolution of the sensor. The radiometric, spatial, spectral, and temporal components of an image or a set of images all provide information that we can use to form interpretations about surface materials and conditions. For each of these properties, we can define the resolution of the images produced by the sensor system. These image resolution factors place limits on what information we can derive from remotely sensed images.

Radiometric resolution or radiometric sensitivity refers to the number of digital levels used to express the data collected by the sensor. In general, the greater the number of levels, the greater the detail of information. The number of levels is normally expressed as the number of binary digits needed to store the value of the maximum level, for example, a radiometric resolution of 1 bit would be 2 levels, 2 bit would be 4 levels, and 8 bit would be 256 levels. The number of levels is often referred to as the Digital Number or DN value. Typically, this ranges from 8 to 14 bits, corresponding to 256–16,384 intensities or "shades" of color, in each band. Many current satellite systems quantize data into 256 levels (8 bits of data in a binary encoding system). The thermal infrared bands of the ASTER sensor are quantized into 4096 levels (12 bits). The more levels that can be recorded, the greater is the radiometric resolution of the sensor system. High radiometric resolution is advantageous when you use a computer to process and analyze the numerical values in the bands of a multispectral image.

Spatial resolution refers to the size of a pixel that is recorded in a raster image — typically pixels may correspond to square areas ranging in side length from 1 to 1000 meters and sometimes up to 10 km. There are many measures of spatial resolution; the most common include the Instantaneous Field of View (IFOV) and the Effective Instantaneous Field of View (EIFOV). The Ikonos-2 satellite was launched in September 1999 and has been delivering commercial data since early 2000. Ikonos is the first of the next generation of high spatial resolution satellites. Ikonos data records four channels of multispectral data at 4 meter resolution and one panchromatic channel with 1 meter resolution. This means that Ikonos was the first commercial satellite to deliver near-photographic high-resolution satellite imagery of anywhere in the world. The ETM sensor onboard Landsat satellite has a spatial resolution of 120 m for the thermal-IR band and 30 m for the six reflective bands. MODIS detects the amount of energy reflected from the Earth's surface at a 1 km resolution twice a day.

Spectral resolution refers to the number of different frequency bands recorded — usually, this is equivalent to the number of sensors carried by the satellite or plane. The finer the spectral resolution, the narrower the wavelength range for a particular channel or band. Many remote sensing systems record energy over several separate wavelength ranges at various spectral resolutions. Landsat images have seven bands, including several in the infra-red spectrum. Landsat 7 ETM is a polar-orbiting 8-band multispectral satellite-borne sensor. The spatial resolution is 30 meters for the visible and near-infrared (bands 1–5 and 7). Resolution for the panchromatic (band 8) is 15 meters, and the thermal infrared (band 6) is 60 meters. To provide even greater spectral resolution, the so-called hyperspectral sensors make measurements in dozens to hundreds of adjacent, narrow wavelength bands (as little as 0.1 μm in width). The MODIS satellites are the highest resolving at 36 bands.

The *temporal resolution* is simply the frequency of flyovers by the satellite or plane and is only relevant in time-series studies or those requiring an averaged or mosaic image. The revisit period of a satellite sensor is usually several days. Therefore, the absolute temporal resolution of a remote sensing system to image the exact same area at the same viewing angle a second time is equal to this period. Geostationary satellites, Terra, and Aqua satellites (MODIS) have high temporal resolution (<24 h to 3 days). Landsat has medium temporal resolution (4–16 days). Low temporal resolution is normally >16 days. Some satellite systems such as SPOT are able to point their sensors to image the same area between different satellite passes separated by periods from one to five days. Thus, the actual temporal resolution of a sensor depends on a variety of factors, including the satellite/sensor capabilities, the swath overlap, and latitude.

1.1.2 *Remote sensing process*

The core of geographic problem solving is the description, explanation, and prediction of geographic patterns and processes. Real-world processes that produce physical or cultural patterns on the landscape are often complex and not usually totally identifiable. The nature of these processes can be described by two general categories of deterministic and probabilistic processes. Deterministic processes create patterns with total certainty since the outcome can be exactly specified with 100% likelihood. Probabilistic processes concern all situations that cannot be determined with complete certainty. Probabilistic processes can be sub-divided into two useful categories of stochastic and random processes. Stochastic process is when the likelihood of outcomes for some processes can be specified as a set of probabilities: somewhere between total certainty and total uncertainty. Random process occurs when probabilities cannot be assigned to possible outcomes — a situation of total uncertainty. Most spatial and temporal processes are neither completely deterministic nor totally random.

Generally speaking, remote sensing works on the principle of the *inverse problem*. While the object or phenomenon of interest (the *state*) may not be directly measured, there exists some other variable that can be measured (the *observation*), which may be related to the object of interest via some (usually mathematical) model. The common analogy given to describe this is trying to determine the type of animal from its footprints. For example, while it is impossible to directly measure temperatures in the upper atmosphere, it is possible to measure the spectral emissions from a known chemical species (such as carbon dioxide) in that region.

The frequency of the emission may then be related to the temperature in that region via various thermodynamic relations.

The remote sensing process involves two stages: data collection and data analysis. Data collection is accomplished through the use of remote sensing devices that record data on photographic film or as digital data. The resulting data provide a synoptic view of a portion of the Earth's surface but require analysis and interpretation in order to provide meaningful information. This can involve either visual interpretation, which has been the predominant mode of air photo interpretation, or digital analysis, which has been widely used in processing satellite imagery. Remote sensing remains both an art and a science because there is often no obvious, well-documented choice of data collection devices or analysis and interpretation techniques that are best suited to a particular applied problem situation.

To understand the relationship of digital image processing to remotely sensed data, one should have a clear concept of the steps involved in the remote sensing process. These steps are illustrated in Figure 1.1.

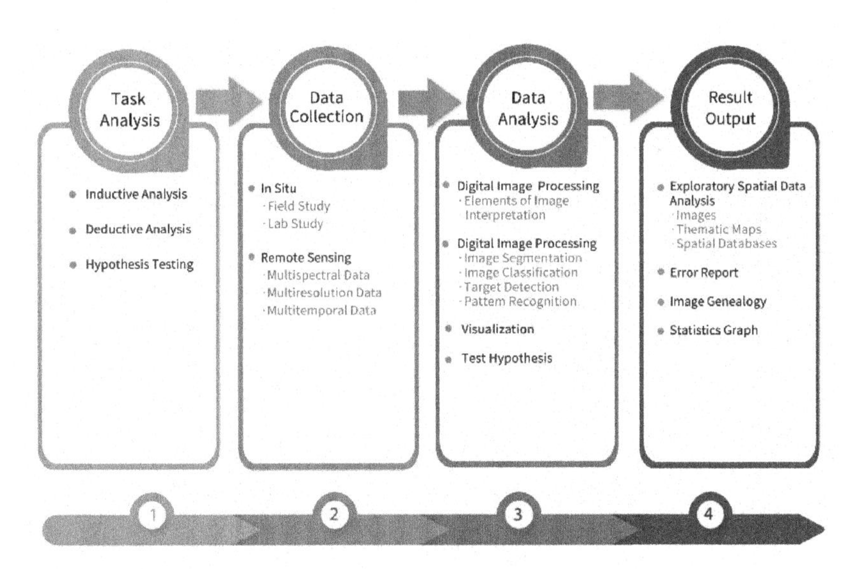

Figure 1.1. The remote sensing process.

There are three basic types of logic that can be applied to a quantitative remote sensing problem: inductive, deductive, and technologic. Scientific approaches use both inductive logic and deductive logic methodologies, while a technological approach uses a technologic logic methodology. The steps in each of these logic methodologies can be seen in Figure 1.2 (Curran, 1987).

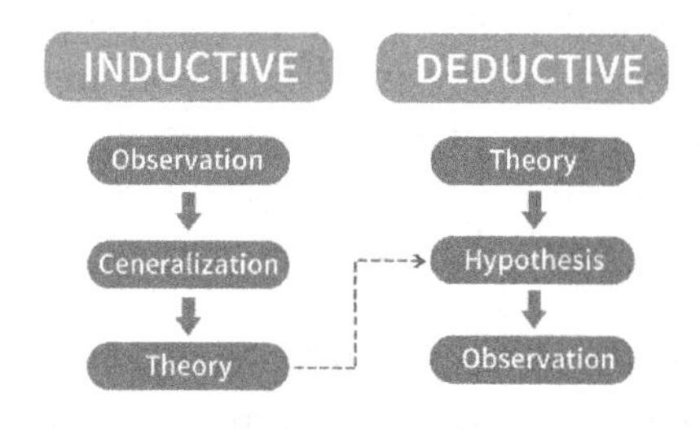

Figure 1.2. Methodologies in remote sensing.

In terms of remote sensing processing, two different fields are digital image processing and information extraction. Digital image processing includes the following:

- Radiometric processing
- Calibration
- Image enhancement
- Atmospheric correction
- Geometric processing
- Geometric characteristics
- Geometric correction
- Image registration
- Image transformation
- Vegetation Indices
- Principal component transformation
- Fourier transformation
- Image classification

The chain of information extraction is as follows:

- Ground receiving station and data ordering
- Radiometric preprocessing (calibration and atmospheric correction)
- Geometric preprocessing
- Feature extraction
- Image classification for category variables
- Estimation of bio-/geo-physical variables
- *In situ* data collection and validation

1.2 Real-Time Remote Sensing Imaging — First Level

A real-time imaging system accepts input in the form of digital data, processes the data, and sends the data to a display device within a given time frame/deadline. Real-time imaging has application in areas such as multimedia, virtual reality, medical imaging, and remote sensing and control. So far, the imaging community has witnessed tremendous growth in research and new ideas in these areas. To lead structure to this growth, Sinha *et al.* (1996) outlined a classification scheme and provide an overview of current research in real-time imaging.

Real-time imaging requires technical support from a multidisciplinary range of research areas, including

- image compression
- advanced computer architecture
- image enhancement and filtering
- remote control and sensing
- computer vision
- optical measurement and inspection.

1.2.1 *Image compression*

Digital images often contain enormous amounts of data, necessitating extensive requirements for storage, process, and communication. The basic task of image compression is to reduce the amount of data required to represent the image. Generally, compression is achieved by removing redundant data and encoding the image by using a suitable coding scheme. The coded data can then be efficiently transmitted or stored, and later decoded for display or processing.

With the increasing importance and technical demands of cyberspace, the need for image compression has grown steadily, and image compression is a major research topic in the area of multimedia computing. This need is intensified by the increasing use of digital computers in video production, animated movies, printing, and publication. Furthermore, image compression plays a crucial role in many diverse applications, including televideo-conferencing, satellite imagery for weather and other earth-resource applications, astronomy, medical imaging, and document processing, as well as in military, space, and hazardous materials' applications.

Compression techniques fall into two groups: lossless (information preserving) and lossy. Lossless techniques, e.g., Huffman encoding, provide perfect, but usually slower, regeneration of the image. Lossy techniques result in higher compression ratios, often faster image reconstruction, but the production of the original image is not perfect. Such techniques are often applied in image and video compression, such as in videoconferencing and facsimile transmission, where small errors in the quality can be tolerated. Lossy techniques can be classified into the following: (1) prediction-based, (2) frequency-based, and (3) importance-oriented, including the Joint Photographic Experts Group (JPEG) and the Motion Picture Experts Group (MPEG) px64 or H26l. Research challenges in this area include developing new real-time compression algorithms (possibly hybrid). Some other evolving compression techniques include wavelet-based compression, fractal-based compression digital video interface (DVI) compression, and sub-band image and video coding. The recent development of wavelet compression can yield "virtual lossless" images with a compression ratio between 10:1 and 20:1, which can reduce the amount of space required to store raster images (Xue and Rees, 1999). A comprehensive survey on various image compression techniques and standards for multimedia applications can be found in Furht (1995a, 1995b, 1996).

1.2.2 *Advanced computer architecture*

Features of computer architectures quite naturally have a significant effect on the execution time of a program and thus on the response time of a real-time system. Parallel processing is employed to speed up computation. Fortunately, the natural structure of digital images and the standard operations on those images are well suited for manipulation by parallel architectures, which make use of pipelining and task and array parallelism to accomplish computationally intensive tasks. In addition, such processors are often configured to increase the throughput and thereby possibly meet the real-time constraints of the image processing system. Wu and Guan (1995) describe requirements for distributed real-time image processing systems.

1.2.3 *Computer vision in remote sensing*

(Real-time) computer vision comprises many techniques used to estimate quantitative knowledge parameters and to enable anal quantitative decisions. Importantly, it requires techniques for geometric modeling and cognitive processing. Moreover, all of this processing must occur in real time; the difference between the time at which an object is sensed and the time when a decision is made must be less than a prespecified deadline (Sinha *et al.*, 1996). Detection of a given object in real time by processing imaging information is a function of various factors (Dougherty and Giardina, 1987): (1) the overall set of objects that might be viewed, (2) their similarity to the target object, (3) the types of sensors involved, (4) the amount of distortion and noise, (5) the algorithms and architectures employed, (6) real-time constraints, and (7) the required level of confidence in the decision.

Wechsler (1993) gives an overview of a taxonomy of parallel architectures and of the relevance of parallel hardware architectures and of the relevance of parallel hardware architecture for computer vision tasks.

1.2.4 *Image enhancement and filtering*

The primary goal of image enhancement techniques is to make an image more suitable for a specific application. There are two main approaches for image enhancement: spatial-domain and frequency-domain approaches (Sinha *et al.*, 1996). The spatial-domain approach is based on the manipulations of the pixel image, whereas the frequency-domain approach is based on modification of the Fourier transform of the image. Spatial filters are most commonly used for real-time image processing for edge enhancement; image enhancement features such as pre-emphasis spatial altering and real-time pixel remapping are also widely used. Optical image processing has also proven effective for image enhancement; this technique includes operations such as morphological low-level nonlinear functions, detection of candidate regions of interest, fusion of correlation outputs to reduce false detection, and feature extraction.

1.3 Real-Time Imaging and Processing in Remote Sensing

Remote sensing in its broadest sense is used to mean "observation at a distance" and this is really all we do when we look at something. In its more restricted technical sense, it is usually used to denote the observation of the surface of the Earth from an aircraft or a spacecraft and this usually, but not always, involves the generation of an image. The processing of a remotely sensed image is an indispensable step in many of the applications of remote sensing in environmental and resource monitoring. Since a raw image contains many distortions and sources of error which may obscure the useful information in it, the image must usually be processed so that the user can use it. Different levels of processing may be used, depending on the nature of the information that is required.

There are several processing procedures that are common to very many different applications of remote sensing and these include the following:

- preprocessing of the image
- Geometrical correction (Figure 1.3)
- Radiative correction (Figure 1.4)
- Image processing
- Enhancement (Figure 1.5)
- Pattern recognition and classification (supervised or non-supervised) (Figure 1.6)
- Automatic mapping
- Data storing, i.e., archiving (building a database).

Of course, data management procedures such as calibration, verification, masking, and data compression are all image processing procedures. Either one or several processing operations need to be done according to the task. Moreover, the processing operations may, or may not, need to be

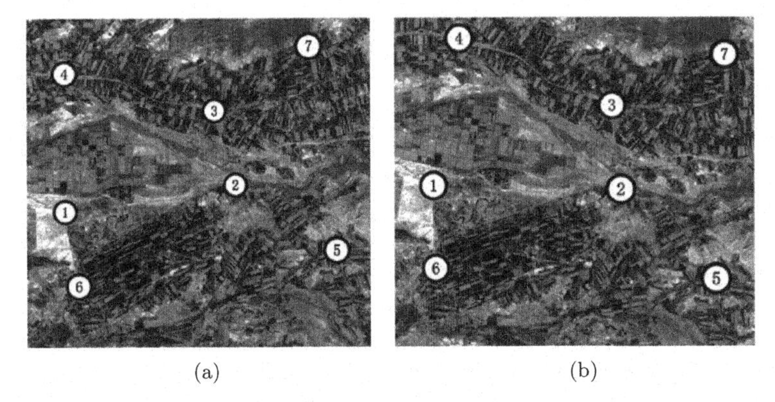

(a) (b)

Figure 1.3. Geometrical correction.

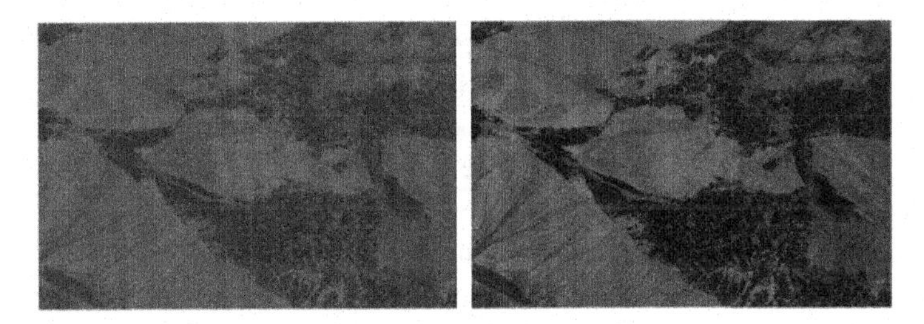

Figure 1.4. Radiative correction.

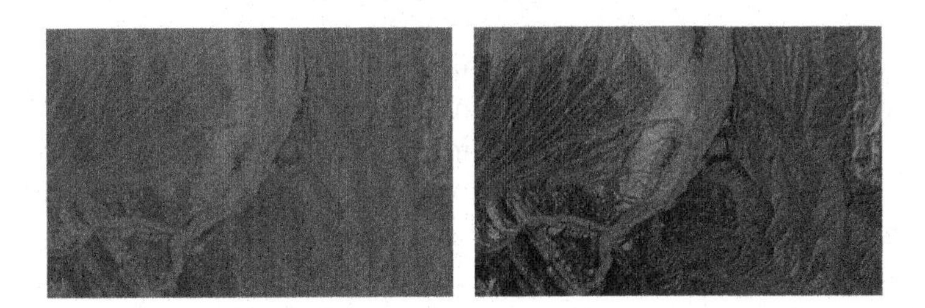

Figure 1.5. Enhancement.

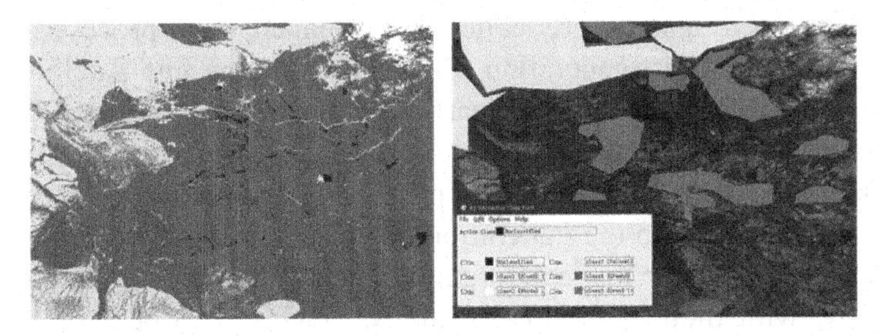

Figure 1.6. Pattern recognition and classification (supervised or non-supervised).

performed in real time, depending on the nature of the application question.

Processing of remote sensing data may be carried out as analog processing or as digital processing. Analog processing is mainly optical processing but there is also some analog electronic processing. The advantages of this method are high speed (for example, an optical Fourier transform will be generated at the speed of light), simple systems, and low cost. The disadvantages are less flexibility, fewer functions (only enhancement, no reorganization), low accuracy (small number of gray levels, only 15 gray levels for a satellite picture, 6–8 levels for human eye, and up to 64 levels for optical instruments), information loss with re-copy, low mapping scale, and always only qualitative analysis is possible. Digital processing is carried out on a remote sensing image processing system with a computer as the main processor and with various input and output formats. Features include the following:

- high accuracy (usually 256 gray levels),
- no information loss with re-copy,
- good consistent, thus large scale mapping,
- parameters are changeable because of programming,
- more functions and more flexibility,
- quantitative analysis.

But digital systems are more complex and low speed (for example, a Fourier transform may take several minutes or longer). The use of digital processing is now very common for image processing of remotely sensed data. It is also possible but relatively uncommon to combine optical processing and digital processing on the same dataset.

The objective of remote sensing is the Earth's surface and the final aim is to apply remotely sensed imagery for environmental and resource monitoring. Advances in sensor technology are revolutionizing the way remotely sensed images are collected, managed, and processed. The incorporation of latest-generation sensors to airborne and satellite platforms is currently producing a nearly continual stream of high dimensional image data, and this explosion in the amount of collected information has rapidly created new processing challenges.

Real time refers to the generation of images or other information to be made available for inspection simultaneously, or very nearly simultaneously, with their acquisition (Sabins, 1987). The "near real-time" concept is used for EO data which are available within a maximum time (~hours) after observation by the satellite. For some problems, perhaps most obviously for geological problems, there is no need whatever for real-time processing of remotely sensed data. However, there are a number of environmental problems such as floods and forest fires for which real-time processing is essential if the remotely sensed data are to be useful in the tackling of an environmental problem rather than just an exercise in academic curiosity. In particular, many current and future applications of remote sensing in Earth science, space science, and soon in planetary exploration science require real or near real-time processing capabilities. Relevant examples include monitoring of natural disasters such as earthquakes and floods, or tracking of man-induced hazards such as wildland and forest fires, oil spills, and other types of chemical contamination. These applications require timely responses for swift decisions which depend upon real-time performance of algorithm analysis. In particular, this is the case for military-oriented reconnaissance and surveillance applications, in which moving targets are of interest, e.g., vehicles in a battlefield, drug trafficking in law enforcement, or chemical and biological agent detection in bio-terrorism.

Due to the diversity and extremely large dimensionality of the image datasets provided by latest-generation observation instruments in Earth and planetary observation applications, there are no commonly accepted real-time image processing architectures and techniques in the context of remote sensing missions. In future years, remote sensing instruments such as hyper-spectral and ultra-spectral imagers will substantially increase their spatial and spectral resolutions (imagers with hundreds of narrow spectral channels are currently available and instruments with thousands of spectral bands are under development), thus producing a nearly continual stream of multidimensional image data. Technological advances are not only expected in optical instruments but also in radar and other types of remote sensing systems. Such explosion in the amount, size, and dimensionality of the information collected on a daily basis presents new challenges for real-time image processing in remote sensing.

We identify three levels of real-time processing (Xue *et al.*, 2002). The first refers to handling the image data directly received from the sensor. This may just involve dumping the image data onto an archive medium and generating a simple quicklook image product. The second level refers to such operations as radiometric calibration, enhancement of the image, geometrical rectification, and cloud screening. The third level involves going all the way to producing an image, chart, map, or data analysis for a specific application.

A special issue on Architectures and Techniques for Real-Time Processing of Remotely Sensed Images from the Journal of Real-Time Image Processing is intended to present the current state-of-the-art and the most recent developments in the task of incorporating (near) real-time image processing techniques and architectures to remote sensing applications (Plaza, 2008). The topics which will be addressed in the special issue include architectures, techniques, and applications for (near) real-time processing of remotely sensed images. In detail, some of them are sensor design for real-time remote sensing applications, real-time onboard processing, and compression of remotely sensed images, VLSI architectures for remotely sensed image processing, parallel and distributed algorithm design and implementation in remote sensing, Grid computing techniques for high-performance remote sensing applications, high-performance heterogeneous, and distributed computing in remote sensing, real-time online processing of remotely sensed images. Combined, these topics will provide an excellent snapshot of the state-of-the-art in those areas and offer a thoughtful perspective of the potential and emerging challenges of applying real-time image processing architectures and systems to realistic remote sensing problems.

Applications of (near) real-time processing of remotely sensed images include efficient content-based image retrieval from remotely sensed image archives, monitoring of natural and man-induced disasters using remotely sensed images, target detection/tracking using remotely sensed images, reconnaissance and surveillance applications, advanced radiative transfer models for real-time remote sensing applications, and real-time weather forecast systems.

1.4 Real-Time Remote Sensing Imaging — First Level

1.4.1 *Airborne remote sensing imaging*

Information travels from the object to the observer usually by means of electro-magnetic waves, such as light, infrared, or radio waves, sometimes by sound. We can detect electro-magnetic waves using various different detectors according to the wavelength of the radiation. One of the earliest recording detectors was the camera, and for well over a hundred years, cameras have been used for what we now describe as remote sensing. Other electromagnetic energy sensors are now built in to various scanners that are currently flown on various airborne and spaceborne platforms to assist in inventorying, mapping, and monitoring Earth resources. Cameras and scanners acquire data on the way various Earth surface features reflect and emit electromagnetic energy and these data are then analyzed to provide information about the resources under investigation.

Aerial photography, as we now know it, was originally developed mainly for military purposes, but nowadays there are a great number of civilian applications, such as aerial photogrammetry (Kraus, 1992), crop condition assessment (Ladouceur *et al*., 1986), and forest reconnaissance and volume estimation (Smith, 1986).

An important feature of remote sensing is the use of multispectral techniques. In these techniques, instead of producing either a black and white panchromatic image, using light of all wavelengths, or a conventional (true) color photograph, what is done is to filter the light into a number of wavelength ranges, or spectral channels, and to generate a set of spatially co-registered monochromatic images, one from each spectral channel. In the early days, the number of spectral channels used was quite small, 4 or 7 or 10, for example. Thus, for each pixel, the signals in the various spectral channel could be used to re-construct a rather wide approximation to the reflectance or emission spectrum of the surface of the Earth. In recent years, the number of spectral channels available has been increased very significantly so that one may now be able to obtain data in several hundred spectral channels that are each very narrow. Thus, for each individual pixel, the signals from all the various spectral channels resemble quite closely a continuous reflection or emission spectrum of the corresponding area on the surface of the Earth. The instrument used then trends to be known as an imaging spectrometer rather than as a multispectral scanner. In recent years, imaging spectrometry has been emerging as a new and promising research area of remote sensing. Imaging spectrometry, which consists of the acquisition of images in many narrow, contiguous spectral bands, generates a hybrid three-dimensional (spatial and spectral) radiance dataset. This leads to difficulties in terms of the presentation of the data. For the early multispectral scanner, e.g., the Landsat Multi-Spectral Scanner (MSS), or Thematic Mapper (TM) or for the NOAA Advanced Very High Resolution Radiometer (AVHRR), it was customary to generate a false color composite image by selecting image data from these channels, assign these images to the three color guns of a color monitor or to produce a corresponding hard copy on paper or as a film transparency. Sometimes, combinations of spectral channels have been chosen to give a presentation of the scene reasonably close to its true colors. Alternatively, to provide the maximum possible information in an image, a principal components analysis has been performed and then, instead of simply choosing these of the spectral bands, the first three principal components were assigned to the three colors of the display system. However, when it comes to the hyper-spectral image data, or three-dimensional data cube, or data cuboid, generated by an imaging spectrometer, it is more difficult to produce adequate visual representations of the data. Such hyper-spectral image datasets allow both that images can be viewed on any given spectral band or band combinations and also that the contiguous spectral radiance curve can be manifested at any given pixel or at any associated targeting area. We mention a few common imaging spectrometers which have been used so far for remote sensing purposes: (1) the Airborne Visible and Infrared Imaging Spectrometer (AVIRIS) developed by JPL (Vane, 1987), (2) the Fluorescence Line Imager (FLI)/Programmable Multispectral Imager (PMI) by Moniteq Ltd. of Toronto (Gower *et al.*, 1985), (3) the Compact Airborne Spectral Imager (CASI) by Itres Ltd. of Calgary Babey and Anger, Borstad *et al.* (1989), and (4) the Reflective Optics System Imaging Spectrometer (ROSIS) by MBB of Ottobrunn in Germany (Kunkel *et al.*, 1991). In addition, a new airborne push broom imaging spectrometer, named the Variable-Interference-Filter Imaging Spectrometer (VIFIS), has been developed at the University of Dundee over the past two years (Sun and Anderson, 1993). As far as real-time processing of the data from airborne instruments is concerned, this has virtually always been only very limited. With cameras, the film is usually processed on the ground after the flight. For scanner, the analog signal output from the detector is usually digitized within the instrument and the digital output is recorded on a magnetic storage medium. The original analog signal or the digitized signal may also be used to generate a quicklook image onboard the aircraft. Beyond this, all further processing of the digital data is usually carried out on the ground afterwards and, therefore, not in real time.

1.4.2 *Satellite remote sensing imaging*

In addition to the airborne instruments, the satellite-flown instruments have been developed rapidly. We have mentioned Landsat MSS and TM and NOAA AVHRR already. In addition to these visible, near-infrared, and thermal infrared mechanical scanners, which have been very widely used, there are other instruments such as the push broom SPOT scanner, the two-look Along Track Scanning Radiometer (ATSR), and a whole variety of instruments using microwave radiation. For the further, there are several planned satellite imaging spectrometers: NASA's High-Resolution Imaging Spectrometer (HIRIS) (Dozier and Goetz, 1989), ESA's Medium Resolution Imaging Spectrometer (MERIS) (Baudin *et al.*, 1990), and NASA's Moderate Resolution Imaging Spectrometer with Tilt (MODIST) (Salomonson *et al.*, 1989).

As far as satellite-flown instruments are concerned, the data are subjected to a certain amount of on-board processing in real time. This is rather limited and is basically concerned with assembling the output data from the on-board instruments, along with some housekeeping and calibration data, into the specified format; the data can then either be transmitted directly (direct broadcast transmission) for reception by a ground station that is within range or recorded on an on-board tape recorder for subsequent playback and broadcast transmission later in the orbit when the spacecraft is within range of a chosen ground station. In the past, it has been the practice of ground stations simply to store the received data in an attempt for any further real-time processing. This philosophy is now changing because of the enormous increase in the speed and power of computing systems. Now, some of the level 2 and level 3 processing, see subsequent sections, is beginning to be performed as soon as the data are received at a ground station.

Many of the scientific instruments on satellites in low Earth orbits, which are used to observe the Earth and its atmospheric processes, are rigidly mounted on the main body of the spacecraft and therefore afford no significant degree of pointing control. Exceptions are the Coastal Zone Color Scanner (CZCS) and SPOT which do have a pointing capability. The AURIO combination of instruments, proposed for flight on a European Polar Platform, makes use of pointing control capabilities to enable the observation and tracking of features in the aurora, which exhibit a wide range of temporal and spatial scales (Shahidi *et al.*, 1991).

Systems designed to provide the marine user with operational environmental information require a means to communicate data to sea in near-real time. The transfer of data (images) obtained by environmental remote sensing satellites is complicated by the raster-based structure of the corresponding data files and by their relatively large size. Analysis of marine mobile communications technologies indicates that the least expensive for this application is cellular telephony (Whitehouse and Landers, 1994). Experiments were conducted off the coast of eastern Canada to determine the usefulness of cellular telephony for transferring remote sensing imagery to the coastal zone. They included sea surface temperature (SST) images derived from NOAA AVHRR data and synthetic aperture radar (SAR) images obtained by ERS-1. Results have demonstrated that cellular telephony can satisfy the technical, time, and cost requirements of coastal operational remote sensing applications. By incorporating data compression techniques, SST images were transmitted at a maximum rate of 12.2 kbps (kilobits per second). Processed SAR images were transmitted at significantly lower rates. Transmissions up to 50 km from shore were completed successfully. Comparison of these results with those obtained in other regions indicates that processed SST images have been transmitted at rates of 21 kbps and to distances as great as 150 km from shore. Investigation of such variations indicates that transmission speed is influenced by the extent to which the image has been processed prior to transmission and by the use of compression techniques. Transmission distance varies from location to location, largely due to variations in regional atmospheric effects and the quality of shipboard installations.

The design of a Danish high-resolution airborne SAR was started in 1986 and the initial test flights took place in November and December 1989. An airborne real-time processor is currently under development, and a first reduced version is expected to be operational in mid-1992. The real-time processor is primarily intended to assist the operator in using the SAR system, but since the processor has been designed to produce high-quality imagery, it is expected to render offline processing superfluous in many cases (Dall *et al.*, 1992). The range-Doppler algorithm is adopted and supplemented with an extensive motion compensation scheme taking into account the special conditions associated with real-time strip mapping of large scenes. From an architectural point of view, the processor is a pipeline of approximately 20 elements interconnected by a dedicated data path and a control bus. Only three different types of elements are involved: a programmable signal processing element, a multipurpose memory element, and a multipurpose interface element. Two commercially available control boards and a display system support the pipeline.

1.5 Real-Time Remote Sensing Processing — Second Level

We defined the second level of processing of remote sensing data as involving such processing as radiometric calibration, geometrical rectification, cloud screening, and simple enhancement of the image. In the early days of remote sensing satellite, the technology was such that ground stations operated under the philosophy that the data received from the satellite was archived and quicklooks were produced. Analysis of the data was regarded as being something to be undertaken by users and not as part of the routine operation of the ground station.

1.5.1 *Calibration*

An accurate calibration is a critical component of any radiometric system (Asmus and Grant, 1999). The SBR system has several different calibration methods built in and at least one must be used before data are collected. Generally, the calibration process must determine the relationship between the radiometer voltage and the actual brightness temperature. In the simplest case, this relationship can be determined by measuring the voltage when pointing the radiometers at a known hot load and then measuring the voltage when pointing the radiometer at a known cold load. These two points are subsequently used to establish a linear relationship. This is called a two-point calibration, and the SBR system will lead the user through the process instructing when the load should be changed and automatically calculating the parameters. The cold load can be a liquid nitrogen-soaked absorber, or it is often simply the sky temperature. If the sky is used, the tracker automatically rotates the radiometer upward to complete the measurement.

Another method often used for calibration is the three-point method. In this method, the user takes a reading at approximately a 60° angle from zenith and then a measurement at zenith. This corresponds to a measurement of two atmospheres and one atmosphere, respectively. From a two and one atmosphere measurement, a zero atmosphere, nearly zero degree voltage, can be evaluated by extrapolation. This accomplishes the establishment of the cold temperature calibration point. As in the two-point method, the hot calibration point is determined by directly measuring the voltage when looking at a known hot load. Again, SBR can automatically lead the user through this type of calibration.

The third method of calibration is the n-point calibration method. This is an extension of the three-point method, whereby several measurements are made between 60° from zenith and 0° from zenith. The best-fit line is determined between these points, and from this, a voltage corresponding to a zero atmosphere brightness temperature can be established. This forms the cold calibration point, and the hot calibration point is again measured in the same fashion as above. Calibrations can be saved in arbitrarily named configuration files to facilitate easy recall and switching if necessary.

Image noise is any unwanted disturbance in an image. If the noise is severe enough to degrade the quality of the image, i.e., to impair the extraction of information from the image, an attempt at noise suppression is warranted. Preliminary analysis of the image and noise structure is usually necessary before applying a noise suppression algorithm. In some cases, sensor calibration data and even test images may exist that are sufficiently comprehensive to estimate noise parameters. Unfortunately, however, these types of data are often incomplete or the noise is unexpected. We are then faced with learning as much as we can about the noise from the noisy image itself. Periodic noise is relatively easy to diagnose and characterize from the imagery, but random noise is considerably more difficult to separate from valid image data. Image noise is divided into several distinct categories such as random noise, isolated noise, stationary periodic noise, and non-stationary periodic noise (Schowengerdt, 1983); however combinations of these types may occur in actual images.

Interpolation lies at the heart of classical numerical analysis. There are two main reasons for this (Ralston, 1965). The first is that in hand computation there is a continual need to look up the value of a function in a table. In order to find the value of the function at non-tabulated arguments, it is necessary to interpolate. Moreover, the highly accurate tables at small increments of the argument that we take for granted today are mostly of comparatively recent origin. Therefore, classical numerical analysts developed an extremely sophisticated group of interpolation methods. Today, the need to interpolate arises comparatively seldom; for example, on digital computers, we almost always generate directly the value of a function rather than interpolate it in a table of values. Moreover, when the need to interpolate in a table does arise, the small increments in the arguments in most tables mean that quite simple techniques (e.g., linear or quadratic interpolation) will usually suffice. Thus, while every numerical analyst must know how to interpolate, he will seldom, if ever, have a use for the more sophisticated interpolation techniques. The second reason is that interpolation formulae are the starting points in the derivation of many methods in other areas of numerical analysis. Almost all the classical methods of numerical differentiation and numerical integration are directly derivable from interpolation formulae. An application of interpolation for image restoration, the moving interpolation window method for noise removal, was developed by Xue (1995). Moving window interpolation is a simple noise-removing method. It can be used to remove random noise and non-random noise, especially the combination of several kinds of noise.

1.5.2 *Navigation of satellite images*

There is now an extensive and very interesting literature on the "navigation" problem of relating AVHRR data to positions on the Earth's surface (see Emery *et al.*, 1989, for a review, which is updated in the introduction to Rosborough *et al.*, 1994, and see also the excellent book *Kidder and Vonder Haar*, 1994). This relationship can be approached in either of two ways (called indirect and direct referencing respectively in Ho and Asem, 1986): (i) the data can be mapped pixel by pixel onto a user-defined map projection of part of the Earth's surface or (ii) the satellite image is left alone and itself determines the map projection which is then calculated as an overlay on the image. In this latter approach, in particular, it is crucial that the software should be able to report the latitude and longitude of any given pixel.

An algorithm was presented by Klaes and Georg (1992) to reprocess the navigation procedure for positioning NOAA-AVHRR pixels on any map projection in an efficient way, allowing the data to be positioned either during ingest or shortly after the end of reception. The preprocessing procedure requires about 10 min on a MicroVAX II computer; the application was demonstrated with the Satellite Data Processing System

(SDPS) of the German Military Geophysical Office (GMGO) by positioning data to a polar stereographic map.

In any event, therefore, we need to be able to go from a pixel to the corresponding latitude and longitude, and this is either used as in (i) above or as the reporting mechanism for the overlay in (ii) above. This section is adapted from Huggett and Opierting mechanautomatic navigation of AVHRR images (Huggett and Opie, 2002). Their method for generating an overlay on a satellite image and reporting latitude and longitude from cursor positions on that image is simple and very accurate without use of ground control points. The navigation is near-real time: within 30 min of first acquisition of the satellite signal, a geo-located HRPT image can be produced in an export format of the user's choice.

The AVHRR data themselves are transmitted in two ways. In the Automatic Picture Transmission (APT) transmission, the scan lines have been "linearized" (see Kone, 1992, for example) so as to compensate for the line of sight foreshortening suffered by the AVHRR, while in the High-Resolution Picture Transmission (HRPT) transmission, the raw data are available. Our method was not only designed for the HRPT transmission but can also cope with APT.

A significant part of the literature is concerned with the prediction of the sub-satellite point for each scan line, using some sort of orbital model. In our case, the SGP4 system is used (see Hoots and Roehrich, 1988), which among other things will return a sub-satellite point and satellite height for any given time, once the orbital elements for that satellite have been specified. Other work (such as Crawford *et al.*, 1996) uses SGP4 but not in such a thoroughgoing way.

1.5.2.1 *The mathematical model*

The method is to convert from geocentric latitude, which we call w, and longitude, θ, to three-dimensional vectors based on the usual orthogonal x, y, z axes with origin at the center of the Earth and x axis intersecting the Greenwich meridian at the equator. Then, all the calculations are done in terms of these vectors, and if necessary, we can convert them back to w and h. First, we describe how to convert from (θ, ϕ) to the corresponding vector $\mathbf{p}$. We refer to this routine as *vect*, defined as follows:

$$\mathbf{p} = vect(\theta,\phi) = (R(\phi)\cos\theta\cos\phi,\ R(\phi)\sin\theta\cos\phi,\ R(\phi)\sin\phi \qquad (1.1)$$

$R(\phi)$ is the radius of the Earth, according to the spheroid defined by WGS72. Now, the inverse of this operation is as follows,[1] where **p** has components (x, y, z):

$$\theta = \sin^{-1}\left(\frac{y}{\sqrt{x^2+y^2+z^2}\cos\phi}\right), \quad \phi = \sin^{-1}\left(\frac{z}{\sqrt{x^2+y^2+z^2}}\right) \tag{1.2}$$

The scan lines are perpendicular to the orbital path of the satellite and are therefore not perpendicular to the ground track because of the rotation of the Earth. The sub-satellite points, however, whose positions come from SGP4, lie along the ground track. So, given a particular sub-satellite point with position vector **e**, our first task is to calculate a vector lying along the scan line at that point. This will in fact be the "normal" vector to the plane of the orbit; call it **u**. Next, we need a vector **v** perpendicular to **u** and tangential to the Earth's surface at our sub-satellite point; **v** will be in the direction of motion of the satellite.

Let us do this in more detail. Let (ϕ_i, λ_i) be the longitude and geodetic latitude of the sub-satellite point of the center of the *i*th scan line, where *i* starts at zero, and let a *i* be the satellite altitude at that point (taken from SGP4 and measured above the oblate Earth). Also, let R_e and R_p be the Earth's radii at the Equator and the Pole, respectively. Then, we convert to geocentric latitude ϕ_i by

$$\phi_i = \tan^{-1}\left[\tan\lambda_i\left(\frac{R_p^2 + a_i\sqrt{R_e^2\cos^2\lambda_i + R_e^2\sin^2\lambda_i}}{R_e^2 + a_i\sqrt{R_e^2\cos^2\lambda_i R_e^2\sin^2\lambda_i}}\right)\right] \tag{1.3}$$

(Borkowski describes this conversion in detail, and there are shorter discussions in NOAA, 1999; Straka *et al.*, 1993.) The vector from the center of the Earth to this sub-satellite point is

$$e_i = vect(\theta_i, \phi_i) \tag{1.4}$$

Now, using the inclination *i* of the orbit which comes from SGP4, we can calculate our vector **u** *i* perpendicular to the orbital plane as follows.

---

[1]As it is written here, the code would only report longitudes less than 90°, so the sign of y ought to be used too.

It makes an angle i with the unit vector $\mathbf{k}$ (in the z direction), and it is perpendicular to $\mathbf{e}_i$. So, it must satisfy

$$\mathbf{u}_\mathrm{i} \cdot \mathbf{k} = \cos i, \qquad \mathbf{u}_\mathrm{i} \cdot \mathbf{e}_\mathrm{i} = 0 \tag{1.5}$$

and the vector $\mathbf{v}_i$ is then given by

$$\mathbf{v}_\mathrm{i} = \mathbf{u}_\mathrm{i} \times \mathbf{e}_\mathrm{i} \tag{1.6}$$

Finally, we adjust the lengths of $\mathbf{u}_i$ and $\mathbf{v}_i$ so that $\mathbf{u}_i$ has length 1 and $\mathbf{v}_i$ has length equal to the width of the ith scan line. Then, once for each image, we construct a look-up table for the subsequent code, consisting of $\mathbf{e}_i$, $\mathbf{v}_i$, $\mathbf{u}_i$, a_i for each i.

1.5.2.2 *Creating the overlay*

We will label the variables on the screen s and t so that we can use pairs (s, t) synonymously with pixels. The s axis is horizontal (think of s for "scan" or "sideways") and the t axis is vertical (think of "time" or "track"). Given the geodetic latitude l and longitude for a particular geographical location, our object is to find the corresponding pair (s, t). We first have to find ϕ and θ, so we have to convert from geodetic to geocentric latitude

$$\theta = \tan^{-1}\left(\frac{R_p^2 \tan\lambda}{R_e^2}\right) \tag{1.7}$$

and then we calculate the vector

$$\mathbf{P} = vect(\theta,\phi) \tag{1.8}$$

Now, for each value of i successively, we calculate the vector from the ith sub-satellite point to the point (ϕ, θ):

$$\mathbf{q} = \mathbf{p}\ \mathbf{e}_\mathrm{i} \tag{1.9}$$

If the projection of the vector $\mathbf{q}$ in the direction of $\mathbf{v}_i$ is less than the length of $\mathbf{v}_i$, then the point (θ, ϕ) must lie in the ith scan line and so we set $t = i$, and we have the scan line containing the specified geographical location.

It remains to determine *s*, its "sideways" pixel number. Let δ be the swathe angle so that the angle away from the vertical of the AVHRR scanner (ψ) varies from δ to δ and let $2n$ be the number of pixels across the screen. The variable s is linearly related to ψ:

$$s = \frac{2n\psi}{\delta} \tag{1.10}$$

Suppose η is the corresponding angle subtended at the center of the Earth (see Figure 1.7).

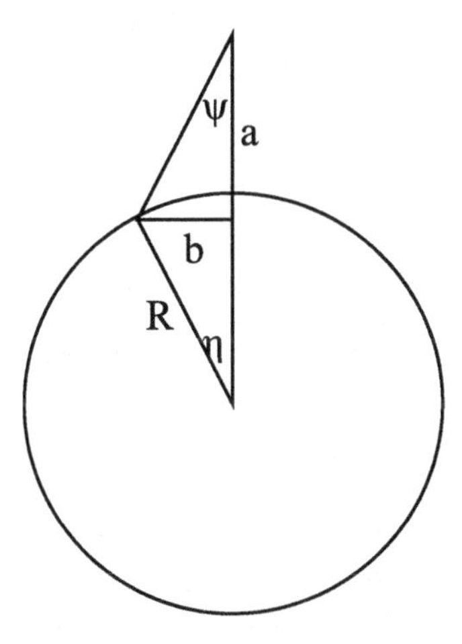

Figure 1.7. Underlying geometry.

Then, we have

$$b(\cot\psi \_+\cot\eta) = a + R, \quad \sin\eta = \frac{b}{R} \tag{1.11}$$

and hence

$$\cot\psi + \cot\eta = \frac{a+R}{b} = \frac{a+R}{R\sin\eta} = k\cos ec\eta \tag{1.12}$$

where $k = 1 + a_t/R_t$. Here, R_t is the radius of the Earth at the latitude ϕ_t. Therefore (as in Brush, 1988; Emery *et al.*, 1989),

$$\psi = \tan^{-1}\left(\frac{\sin\eta}{k-\cos\eta}\right) \tag{1.13}$$

So, we need to find η, which can easily be calculated from

$$\cos\eta = \frac{p \cdot e_i}{pe_i} \tag{1.14}$$

We now have a method for getting from the latitude and longitude of a geographical location to a pixel on the satellite image. Given a particular satellite image, one could now hunt through a database of locations on coastlines, marking these locations on the image and then interpolating. This in effect draws a coastline, or any other desired outline, as an overlay on the image.

1.5.2.3 *Reporting the geographical position of a pixel*

The procedures in this section can be used in either of the two ways. We will present them as though we already have a satellite image with an overlay on it created as in the previous section, and we wish the software to report the latitude and longitude of the cursor position as it moves over the image. They could be used instead to map the data from the satellite onto a user-defined map projection of part of the Earth's surface. In any event, the problem is as follows: given the position (s, t) of a pixel, we have to calculate (θ, ϕ).

We first calculate the variable g, which is the angle subtended at the center of the Earth, from Equation (1.12). This requires an iteration because the equation uses the Earth's radius, which itself depends on the latitude w. Solving it for g we obtain a quadratic, with root:

$$\cot\eta = \frac{-\cot\psi - \sqrt{k^2 \cos ec^2\psi - k^4}}{1-k^2} \tag{1.15}$$

If $\cot\psi<0$, the geometry requires us to use the other root. Then, put

$$\mathrm{p} = \mathrm{e}_i + R\tan\eta\ \mathrm{u}_t \tag{1.16}$$

Now, we find the longitude and latitude of the vector **p** from Equation (1.2), and we then have the required θ and ϕ. Lastly, we recall that θ is the geocentric latitude, which we have to convert to the geodetic latitude using the inverse of Equation (1.7):

$$\lambda = \tan^{-1}\left(\frac{R_e^2 \tan\phi}{R_p^2}\right) \tag{1.17}$$

Note that the iterative procedure leading to Equation (1.15) replaces that described in Borkowski (1989).

Let us briefly compare the procedure given in NOAA (1999) for reporting latitude and longitude, which is broadly equivalent to ours. There the input data are time, satellite position, satellite velocity, and scan angle, and the coordinate system is fixed in space, not on the Earth. Their method proceeds by determining the vector from the satellite to the scan spot, deducing the coordinates of the scan spot, calculating which point on the Earth's surface was at that spot at that time, and finally transforming to geodetic latitude and longitude as above.

1.5.2.4 *Modifying the procedures for APT*

To create the overlay for an APT image, we first note that the calculation of t is as in the HRPT case. Then, we calculate the angle χ_t subtended at the center of the Earth by the tth scan line, as follows:

$$\chi_t = \cos^{-1}\left[\frac{(a_t + R_t)\sin^2\delta + \cos\delta(R_t^2 - (a_t + R_t)^2\sin^2\delta)^{1/2}}{R_t}\right] \tag{1.18}$$

Then, $R_t\chi_t$ is the length of the scan line so that

$$s = \frac{n}{R_t\chi_t}\mathrm{q}\cdot\mathrm{u}_t \tag{1.19}$$

where $\mathbf{q}\cdot\mathbf{u}_t$ is the projection of the vector $\mathbf{q}$ (found using Equation (1.9)) in the direction of $\mathbf{u}_t$.

The procedure for reporting the geographical position of a pixel is simpler for APT: we calculate

$$\sigma = R_t \tan\eta \tag{1.20}$$

where $\eta = s\chi_t/n$ is the angle subtended at the center of the Earth. Then, use Equation (1.16) as before.

1.5.2.5 *Timings and outputs*

The interest has been in the development of a fast and accurate navigation system well suited to the manual, semi-automatic, and fully automatic re-projection of large area imagery such as HRPT. This emphasis arises because we are dealing essentially with the requirement for near-real-time processing of such imagery. Our primary concern is therefore the capture and processing of these transmissions so as to provide ready-made navigated image data for well-proven image-processing packages such as ERDAS, PCI, ENVI, or ER-mapper. The multiband nature of these packages means that we can export projected data with raw 10-bit, calibrated, and product layers. Histogram analysis can also be pre-prepared.

The algorithm described above is both fast and accurate but clearly depends on the Earth model being used for re-projection: SGP4 uses WGS72 (see Snyder, 1987). When overlaying onto the image vectors based on latitude and longitude, such as coastlines, we have consistently used WGS72 and the E005 ellipsoid. It is significant that even with an image of 1200 km by 1200 km, which would cover the entire Spanish peninsula, the errors are no more than one pixel (1.1 km) within 50° of nadir and no more than two pixels at the extreme edges where bottleneck correction has been the greatest.

More important is the time it takes to achieve this accuracy in comparison with standard HRPT import and GCP correction methods available elsewhere. Within 15 min of the end of real-time acquisition of an HRPT image (which is about 100 Mb of data), it is possible to perform data and product analysis on an image-processing station using native format files that are geo-located and fully calibrated. This includes the production of 11 additional bands consisting of calibrated data, solar and sensor angles, and NDVI, MCSST, and LST products, making 16 bands in total (see Table 1.1).

Moreover, the product bands such as NDVI are formula-driven, with the formulae being user-editable and therefore compiled and edited at run-time. The excellent timing is therefore not a result of specifically optimized code: the process is entirely generic.

The above approach has been found to work extremely well for both HRPT and APT transmissions, provided care is taken over the choice of Earth model and ellipsoid. This became evident when preparing images for the British Antarctic Survey (BAS): using generalized World Data Base II vectors for the coastlines gave appalling results. When we

Table 1.1. The 16 exported channels.

Band	HRPT channel	Content	Formula
1	1	DN sensor count	–
2	2	DN sensor count	–
3	3	DN sensor count	–
4	4	DN sensor count	–
5	5	DN sensor count	–
6	Derived band	Albedo % — HRPT ch1	Internal calibration
7	Derived band	Albedo % — HRPT ch2	Internal calibration
8	Derived band	Temperature °C — HRPT ch3	Internal calibration
9	Derived band	Temperature °C — HRPT ch4	Internal calibration
10	Derived band	Temperature °C — HRPT ch5	Internal calibration
11	Derived band	Solar azimuth angles — degrees	Internal
12	Derived band	Solar elevation angles — degrees	Internal
13	Derived band	Pixel-nadir angles — degrees	Internal
14	Derived band	NDVI (ratio)	User defined
15	Derived band	SST °C	User defined
16	Derived band	LST °C	User defined

replaced them with a set of vectors specially created and corrected for polar studies by the BAS, the accuracy, bearing in mind the ever-changing coastlines of these regions, was impressive.

For the APT transmissions, some averaging and scaling takes place before the transmission of the signal (see NOAA, 1979), whereby a total of 909 of the original 2048 HRPT pixels per line are used to create the APT version. The "index of cooperation" is a fundamental aspect of this process, as shown in Figure 4 in Joerg Kilngenfuss Pub. (1987). Space and calibration telemetry data and synchronization signals are added to either side of the APT image, creating a final analog line signal generated from a total of 1040 digital samples. Navigation problems arise therefore because any particular APT collection and storage system will digitize the incoming analog signal in its own specific way. For example, some systems will maintain spatial integrity, digitizing all the original pixels, while others will simply take "*n*" samples over the "line" period.

Two ways of dealing with this are (i) to introduce a scaling factor into the algorithm for APT which will modify the apparent width of the APT image line or (ii) simply to use a modified value for the swathe angle. The effect will be identical. We have chosen the latter method, altering the swathe angle from the published value of $1024 \times 0.0541 = 55.3984$ (NOAA, 1999) for AVHRR transmissions prior to NOAA-16.

Another source of errors that Huggett and Opie (2002) have found to be important lies within the SGP4 model itself: it suffers from a significant deterioration in positional accuracy over a 30-day period. Once the orbital data are between two and three weeks old, our studies show that errors on the ground can increase to between 5 and 8 pixels (almost 10 km) over the whole image. Other models, such as JGM and HPOP, can reduce this. We do not at present use these, relying instead on accurate timing from GPS units and up-to-date orbital data loaded automatically by our acquisition computer via the Internet.

It is worth comparing this technique (referred to as the analytical approach in Campbell, 1996) with the resampling methods used in particular for Landsat images. The latter rely on local algorithms such as cubic convolution to perform the geometric correction, and they are less accurate by far on large HRPT scale imagery and at least three times more time-consuming, even assuming that an optimum number of accurate and well-positioned ground control points can be identified. In addition, the overall spectral qualities of the corrected image will differ from those of the data, as a result of the averaging which takes place. Our algorithm suffers from none of these drawbacks, being a relatively simple and robust non-local geometrical one designed for images on the HRPT scale. It is fast, accurate to within two pixels (2.2 km) over the whole image, and accurate to around one pixel (1.1 km) for data less than 50° from the nadir.

There are other possibilities for improvement: instead of simply using the outputs from SGP4, it may be possible to access its internal data structures, thus streamlining our whole algorithm. Future developments would also need to incorporate the specifications for the LRPT scanners. Perhaps most significantly, though, the accuracy of this procedure already makes a fully automated image processing algorithm feasible.

1.5.3 *Fast delivery of remotely sensed images*

In the past, users were normally supplied with raw data. However, the philosophy of ground station operation is now changing for several reasons: (i) the practical consequence of the former philosophy was that large quantities of useful data were never exploited at all, (ii) the power and speed of computers have been totally transformed over the last one or two decades, and (iii) it is more efficient to calculate a product such as a chart of sea surface temperature or vegetation index on a regular routine basis once for all from the data from each satellite pass rather than have several different users all processing the data separately to produce the same information from the same dataset. Thus, the present philosophy of running a ground station is now much more to process the data to generate information or products in a form that is needed by users rather than simply archiving the raw data and supplying the users with raw data when requested. The philosophy is that users want information and they generally do not want raw data unless they are concerned with quite basic research.

For some applications, perhaps most notably geological studies, there is no particular need for the remotely sensed data to be processed in real time. However, even in this situation, remote sensing users often require access to the data without undue delay. In some applications, however, it is essential that the data are processed in near-real time, otherwise the information is useless. One obvious example is weather forecasting. Other examples are provided by disaster monitoring situations.

Remote sensing data are the large-volume data source. Historically, locating and browsing satellite data have been a cumbersome and expensive process. This has impeded the efficient and effective use of satellite data in geosciences. This may be in near to real time or after browsing an archive or a catalog system.

In the past, remote users of the archive held at a satellite receiving station have had to select from an online catalog of the recorded satellite datasets stating their general area of interest (Soukeras and Hayes, 1991). The drawback to the catalog system was that it provided only a very crude means of identifying available data without indication of imaging quality, cloud cover, or other pictorial cues of image suitability. The result is that the remote end user did not always have an estimation of the suitability of a particular dataset before ordering or downloading the data.

As an example of the change in philosophy of ground station operation, we mention that in the NOAA HRPT data received by the ground station in Dundee are now available over the UK Joint Academic Network (JANET) and the international INTERNET by file transfer from Dundee (NERC Satellite Station, 1993). This facility allows near-real time access to full-resolution HRPT data received at Dundee by any user with the appreciative network connections and avoids the usual processing and postal delays. A number of other network data services are available from Dundee including an online digital quicklook archive and a NOAA pass database.

There is a new interactive tool SSABLE for the archive, browse, order, and distribution of satellite date based upon X Windows, high bandwidth networks, and digital image rendering techniques (Simpson and Harkins, 1993). SSABLE provides for automatically constructing relational database queries to archive image datasets based on time, date, geographical location, and other selection criteria. SSABLE also provides a visual representation of the selected archive data for viewing on the user's X terminal. SSABLE is a near-real-time system; for example, data were added by Simpson and Harkins (1993) to SSABLE's database within 10 min after capture. SSABLE is network-independent and machine-independent; it will run identically on any machine which satisfies the following three requirements: (1) has a bit-mapped display (monochrome or greater), (2) is running the X Window system, and (3) is on a network directly reachable by the SSABLE system. SSABLE has been evaluated at over 100 international sites. Network response time in the United States and Canada varies between 4 and 7 s for browse image updates; reported transmission times to Europe and Australia typically are 20–25 s.

1.6 Real-Time Remote Sensing Image Analysis — Third Level

We defined the third level of real-time processing as involving going to the final stage of producing a chart, map, or processed image for a specific application. One of the very common ways to use remotely sensed data is in a multispectral supervised or unsupervised classification. This may be for a simple cartographic purpose or it may be for something such as a study of land use or of land cover. Such applications, however, do not usually need the processing to be done in real time. In this section, we concentrate on some examples of real-time image analysis and real-time mapping in remote sensing.

One of the first disciplines in which remote sensing has had obvious applications is in meteorology. Meteorological applications require the data in near-real time (Saunders and Seguin, 1992). The Air Force Global Weather Central (AFGWC) Real-Time Nephanalysis (RTNEPH) is an automated cloud model that produces a 48-km gridded analysis of cloud amount, cloud type, and cloud height (Hamill *et al*., 1992). Its primary input is imagery from polar-orbiting satellites. CLAVR (cloud from AVHRR (Advanced Very High Resolution Radiometer)) is a global cloud dataset under development at NOAA/NESDIS (National Environmental Satellite, Data, and Information Service). Only CLAVR and RTNEPH can satisfy the real time requirements for numerical weather prediction (NWP) models with intercomparison of CLAVR, the US Air Force Real-Time Nephanalysis (RTNEPH), and the International Satellite Cloud Climatology Project (ISCCP) by Hou *et al.* (1993).

There is a growing demand for global ozone measurements from satellite instruments for various purposes. For example, an interesting recent application is the assimilation of satellite ozone observations in Numerical Weather Prediction (NWP) models. This is expected to give a better description of the stratospheric dynamics, the temperatures, and the radiation and heating in the NWP model (Stoffelen and Eskes, 1999). Satellite ozone data are also used for monitoring the status of the ozone layer (e.g., during the development of the yearly Antarctic ozone hole) and to improve — in combination with data assimilation — the regional and global forecasts of surface UV radiation (the derivation of surface UV from satellite measurements is described in Ziemke *et al.* (2000), and references therein). Another application is the use of satellite ozone data for planning and support of ground-based atmospheric (chemistry) experiments during measurement campaigns. Recent examples are the THESEO and SOLVE campaigns that took place in the 1999–2000 winter in the Arctic (Newman *et al*., 2001). For these kinds of applications, satellite ozone observations are especially useful when they are available in near-real time, i.e., within 3 h after observation. However, the usability of near-real-time satellite ozone data depends on the accuracy of the data and the accuracy requirements of the particular application. Therefore, it is necessary to establish the quality of near-real-time satellite ozone data by validation with other (independent) ozone measurements.

The transmission of fast-delivery (FD) Earth-observing satellite images to ships in order to assist in the direction of oceanographic cruises is nothing new — in fact, the use of sea-surface temperature images has had a remarkable impact in improving the effectiveness of many experiments. However, other sensors have been less valuable in this regard; and until the launching of ERS-1, it was not possible to receive using synthetic aperture (SAR) in such a way (Sherwin *et al.,* 1995).

Valks *et al.* (2002) described a Fast Delivery (FD) system, which retrieves ozone columns in near-real-time from radiation measurements by the Global Ozone Monitoring Experiment (GOME) onboard the European Space Agency's (ESA) second European Remote Sensing (ERS-2) satellite. Due to the near-real-time constraint, the FD system uses only a small part of the raw GOME data that are available in near-real-time.

The FD system performs the following three main functions: data acquisition, data processing, and data distribution. The processing function consists of a level 0–1 processor, a level 1–2 processor, and a level 2–4 processor. The EGOI (Extracted GOME Instrument Header data) files and auxiliary data (stored within the processing environment) are input for the level 0–1 processor. The output of the level 0–1 processor (calibrated reflectance) is input for the level 1–2 processor (together with auxiliary data). The outputs of the level 1–2 processor are level 2 data files containing geo-locations, observation times, and ozone vertical columns with uncertainties. The level 0–1 and 1–2 processors are described in more detail in the next section. In the level 2–4 processing, the ozone columns are assimilated in a three-dimensional tracer transport model and a (level 4) global ozone field is calculated (Jeuken *et al.*, 1999; Eskes *et al.*, 2001). The performance of the data acquisition and processing functions is checked by a monitoring function. At the end, the distribution function takes care of the distribution of the level 2 and 4 data to the end users (by Internet/FTP) and to the Fast Delivery Internet/WWW site, where they are made available for the general public.

The near-real-time constraint is an important aspect of the FD system, which means that the period between the actual GOME measurement and the delivery of the total ozone columns to the end users must be less than 3 h. Since the raw GOME data are downlinked to the ESA ground stations once every orbit, the time delay between the GOME observation and the reception at the ground station can be 100 min at maximum. The collection of the raw data at the ground stations, the creation of the EGOI files, and the transfer of these files to KNMI (Royal Netherlands Meteorological Institute), via ESRIN (ESA's data exploitation center in Frascati, Italy), takes about 75 min. The processing of the EGOI data to the level 2 ozone data and the distribution of the level 2 data to the end users take only a few minutes. Generally, more than 93% of the ozone columns are available and distributed to the end users within 3 h after observation.

The near-real-time ozone columns are in reasonable agreement with the ground-based measurements (within 5% for low and mid-latitudes and within Š10 to +5% for high latitudes) and capture temporal and spatial structures of the ozone field similarly. The level of quality meets the requirements for the discussed near-real-time applications (Stoffelen and Eskes, 1999) and is comparable to that of the online GOME Data Processor (GDP) or the Earth Probe Total Ozone Mapping Spectrometer (EP/TOMS) ozone sensor (Lambert *et al.*, 1999, 2000).

A newly developed remote and real-time method for estimating transpiration rates and conductance of crop canopies was examined in comparison with concurrent measurements by the stem flow gauge method (Inoue *et al.*, 1994). The remote method is based on the energy balance of a plant canopy, which uses the net radiation absorbed by the canopy and the remotely sensed canopy temperature as key inputs. The results indicated that the remote method provides reasonable estimates of canopy transpiration and conductance over a wide range of soil water status and micrometeorological conditions on an instantaneous or daily basis. The method allows continuous measurements of these canopy parameters under natural conditions without either plant sampling or disturbing the micrometeorological environment of the canopy.

A physically based rainfall forecasting model for real-time hydrological applications was developed with emphasis on the utilization of remote sensing observations (French and Krajewski, 1994). Temporal and spatial scales of interest are lead times of the order of hours and areas of the order of 10 km^2. The dynamic model is derived from conservation of mass in a cloud column as defined by the continuity equations for air, liquid water, water vapor, and cloud water. Conservation of momentum is modeled using a semi-Lagrangian frame of reference. The model state is vertically integrated liquid water content in a column of the atmosphere. Additionally, the laws of thermodynamics, adiabatic air parcel theory, and cloud microphysics are applied to derive a basic parametrization of the governing equations of model dynamics. The parametrization is in terms of hydrometeorological observable including radar reflectivity, satellite-infrared brightness temperature, and ground-level air temperature, dew point temperature, and pressure.

The performance of a real-time physically based rainfall forecasting model was examined using radar, satellite, and ground station data for a region of Oklahoma (French *et al.*, 1994). Spatially distributed radar reflectivity observations are coupled with model physics and uncertainty analysis through (1) linearization of model dynamics and (2) a Kalman filter formulation. Operationally available remote sensing observations from radar and satellite, and surface meteorological stations define boundary conditions of the two-dimensional rainfall model. The spatially distributed rainfall was represented by a two-dimensional field of cloud columns and model physics define the evolution of vertically integrated liquid water content (the model state) in space and time.

Rainfall forecasts were evaluated using least squares criteria such as mean error of forecasted rainfall intensity, root mean square error of forecasted rainfall intensity, and correlation coefficient between spatially distributed forecasted and observed rainfall rates. The model performs well compared with two alternative real-time forecasting strategies: persistence and advection forecasting.

Advances in remote sensing from ground-based and spaceborne systems, expanded *in situ* observation networks, and increased low-cost computer capability will allow an unprecedented view of mesoscale weather systems from the local weather office. However, the volume of data from these new instruments, the unconventional quantities measured, and the need for a frequent operational cycle require development of systems to translate this information into products aimed specifically at aiding the forecaster in 0–6 h prediction. In northeast Colorado, an observing network now exists that is similar to those that a local weather office may see within 5–7 years. With GOES and TIROS satellites, Doppler radar, wind profilers, and surface Mesonet stations, a unique opportunity exists to explore the use of such data in forecasting weather phenomena. The scheme, called the Local Analysis and Prediction System (LAPS), objectively analyzes data on a high-resolution, three-dimensional grid. The analyzed fields are used to generate mesoscale forecast products aimed at specific local forecast problems (McGinley *et al.*, 1991). An experiment conducted in the summer of 1989 sought to test the use of a pre-convective index on the difficult problem of convective rain forecasting. The index was configured from surface-based lifted index and kinematically diagnosed vertical motion. The index involved a number of LAPS-derived meteorological fields and the results of the test measured in some sense the quality of those fields. Using radar reflectivity to verify the occurrence or non-occurrence of convective precipitation, forecasts were issued for three time periods on each of 62 exercise days. The results indicated that the index was significantly better than persistence over a range of echo intensities. Skill scores computed from contingency tables indicated that the index had substantial skill in forecasting light convective precipitation with a 1–3 h lead time. Less skill was shown for heavier convective showers. The skill of the index did not depend strongly on the density of surface data but was negatively influenced by mountainous terrain.

Sea ice presents a serious impediment to both shipping and offshore operations in the polar regions. Since sea ice conditions can change within a matter of hours, near-real-time monitoring is required. Airborne data are available in some areas, but collection is expensive and coverage is limited. Satellite images can provide wider coverage, but cloud cover, darkness, and the difficulties of rapid processing and dissemination can limit their use. Information on sea ice cover over longer periods is needed for global climate monitoring. Microwave sensors provide the most practical means of monitoring global sea ice cover since they can operate both at night and day and observe through clouds. Previous studies have concentrated on the use of passive microwave data. Laxon (1994) reviewed the processing adopted at the U.K. Earth Observation Data Centre (EODC) and discussed the routine monitoring of sea ice using the ERS-1 radar altimeter. The low data rate and somewhat simple nature of the data lend themselves to the mapping of global sea ice cover and to operational applications.

Northern hemisphere snow cover area has been operationally mapped by NOAA/NESDIS since November 1966 (Ropelewski, 1991). Digitized snow cover charts have been available since 1982.

Cloud types, cloud patterns, and their evolution are the best indicators of atmospheric processes, ranging from synoptic-scale disturbances to small-scale convection. One of the principal missions of weather satellites is to provide this information in picture form. The latest generation of weather satellites gives data with high spatial and cloud top temperature resolution. An automated multispectral cloud and precipitation classification technique for the analysis of satellite imagery was reported by Liljas (1987). The study was carried out for AVHRR onboard NOAA polar-orbiting satellite. The method, involved a cloud brightness and texture discriminant analysis technique. The basis of the technique is that different cloud types and land and water surfaces have different radiational properties in different parts of the electromagnetic spectrum. Several weather situations from a summer period were tested with the same threshold values in the classification. The method was found to give a correct indication of cloud types and, compared with synoptic observations, much better information concerning cloud distribution. In an operational system, the classification scheme can be built up with "boxes" suited to different weather types. For instance, a threshold can divide cumulonimbus into two parts: one with high probability of thunder and the other with low probability. In a weather situation with only convective clouds, these can form different boxes. Separation of different visible-brightness values can also be provided in order to help forecast the dissipation of morning stratus or fog.

The cloud classified images provide an opportunity for identifying different classes of mesoscale cloud systems so that, if the forecaster is familiar with the topography and its effects, he/she can determine if and how the cloud systems are constrained by surface effects. The color images also give a good indication of the cloud structure in various systems, their stage of development, and also the stability of the atmosphere. The mapping of sea, land ice, sea ice, and clouds from Advanced Very High Resolution Radiometer (AVHRR) images taken over Antarctica in daylight was investigated and a classification scheme was proposed on the basis of thresholds retrieved from multispectral patterns of representative data by Zibordi and Meloni (1991). The scheme, which can be used for real-time analysis of AVHRR images in scientific and logistic activities, gave satisfactory separation of different categories. Major misclassification occurs between ice clouds and land ice because of their very similar spectral signature in the AVHRR channels. Comparison of classified samples, obtained from visual inspection of images and from application of the scheme, exhibits a confusion matrix with accuracy A = 92% over areas almost free from ice clouds.

Geochemical monitoring of the environment using modern analytical methods provides rapid information concerning the distribution of pollutants over large areas. In relation to this, complex ecological investigations of considerable importance are being conducted onboard helicopters, cars, and other vehicles equipped with remote sensing devices which provide real-time determinations of air and ground pollution. Examples of rapid ecological surveys in the St. Petersburg region were given by Mashyanov and Reshetov (1995).

Since August 1988, after a development period of three years, the FAO Remote Sensing Centre has been operating the African Real-Time Environmental Monitoring Information System (ARTEMIS) in support of the Global Information and Early Warning System on Food and Agriculture and the Desert Locust Plague Prevention Programme of FAG (Hielkema and Snijders, 1994). The ARTEMIS system was implemented by FAO in close co-operation with the NASA Goddard Space Flight Center, U.S.A., the National Aerospace Laboratory of the Netherlands, and the University of Reading, U.K. ARTEMIS is a highly automated data acquisition, preprocessing and thematic processing, production and archiving system for real-time precipitation assessment and near-real-time vegetation condition monitoring of Africa, the Near East, and southwest Asia, based on hourly Meteosat thermal infrared and NOAA AVHRR data. The vegetation condition assessment capability is currently being expanded to include the rest of Asia and Latin America. ARTEMIS data products, generated by the system on a 10-day and monthly basis, are currently used operationally by a variety of users at FAO Headquarters and by regional and national food security early warning systems in 16 eastern and southern African countries. The ARTEMIS system plays an important role in the generation and archiving of global satellite-derived environmental datasets for use by FAO and other organizations with global monitoring and assessment mandates. Jointly with the European Space Agency, FAO has been implementing during 1989–1992 a dedicated satellite communications system, Direct Information Access Network for Africa (DIANA), which allows real-time transmission of high volume ARTEMIS digital products to user terminals in Nairobi (Kenya), Accra (Ghana), and Harare (Zimbabwe). The DIANA system, which operates through the Intelsat satellite over the Indian Ocean and the Italian Intelsat ground station of Telespazio in Fucino, is currently being tested and demonstrated for a wide variety of applications of an operational, technical, and administrative nature.

In 1992, growers of high-value crops in central Wisconsin expressed an interest in using real-time remote sensing data for detecting crop stress. In response to this request, the Space Remote Sensing Centre of Stennis Space Center, Mississippi, built a system based on work carried out by the USDA, ARS in Weslaco, Texas (Pearson *et al.*, 1994). This new system included a modification from analog to digital cameras that provided a means of overcoming previous barriers that inhibited real-time information delivery. These barriers include automated band to band registration, data calibration, and automated identification and retrieval of desired data frames.

During the 1992 growing season, weekly airborne acquisitions were attempted with 24–48 h information delivery to the growers. Digital images were acquired for high-value crops such as potatoes, sweet corn, cranberries, and peas. The system enabled detection of stresses such as freeze damage, irrigation problems, and Colorado potato beetle infestations. Experience gained during 1992 guided modification to the system to prepare for the 1993 growing season. This overall effort is helping prepare the market and technology for a real-time agriculture monitoring satellite.

Spaceborne remote sensing of inland water quality is based on the assumption that the relation between the reflectance and the concentration of relevant water quality constituents is known *a priori*. Simultaneous measurements of the upwelling and downwelling irradiance, along with phytoplankton chlorophyll-a, suspended matter, and dissolved organic matter concentration at over 20 water bodies throughout the former USSR, Hungary, Germany, and Bulgaria, were reported by Gitelson *et al.* (1993). The measurements cover different trophic states of water bodies, from oligotrophic to hypertrophic, and different climatic conditions. The results are used to develop an appropriate methodology for the monitoring of eutrophication process in inland waters and to test concepts of inland water quality monitoring from space.

In addition to conventional meteorological techniques, acoustic remote sensing (sodar) has also been found to be a potential technique to study sea breeze circulation. Results of studies of sea breeze circulation on the Indian coastline using acoustic remote sensing were presented by Singal (1991). These results have been found to conform to observations made by conventional techniques. Sodar is a simple, cost-effective, and dependable technique and provides continuous and real time data on onset time, duration, and depth of the internal boundary layer.

1.7 Further Development

Up to now, we have briefly described real-time imaging and mapping in remote sensing. Remote sensing data can also provide a data source for geographical information systems (GISs). A GIS is a system designed to store, manipulate, and display georeferenced data. But, unfortunately, with the limitations of current computer sciences, the GIS can only be used as an assistant for remote sensing data analysis besides its other functions in geosciences. Cracknell (1988) discussed the ways in which remote sensing can contribute towards geographical information systems.

The overall development trend of satellite remote sensing system is satellite design from hardware definition to software definition, satellite function from single to direct and remote integration, and application mode from product-driven to task-driven and event perception, the management mode from separation of measurement, control, operation, and control to the integration of measurement, operation and control services, and the intelligence level from single-satellite intelligence to multisatellite collaborative group intelligence.

The purpose of the paper by Brivio *et al.* (1992) was to describe a procedure for the automatic selection of ground control-points in remote-sensing images of high-relief terrain for alignment with a reference map. This problem has been found to be of strategic importance whenever remote sensing images have to be integrated into a geographical information system (GIS) and processed in real time. The procedure described was based on the study of shadow structures in the satellite image and on their comparison with the computer-generated shadows obtained from the Digital Terrain Model (DTM) of the region. The procedure was developed for a Landsat TM image of the Aurina Valley (in the Pusteresi Alps) with the DTM obtained from an IGM (Istituto Geografico Militare) 1:25,000 reference map, but with minor changes, it can be extended to other remote-sensing images. Comparison of the shadow structures is performed by similarity evaluation of a simplified model of their shapes described by means of inertia ellipses. Each pair of shadow structures, recognized as similar and meeting a number of positional constraints, yields a pair of corresponding points whose coordinates provide input values for determining the parameters in the transformation of the input image into a planometrically corrected one. The performance and robustness of the method and its boundary applicability are assessed. An example is given in which the automatically determined control points are directly inserted in a warping function, with reasonably good results.

When considering the research needs related to the integration of remote sensing information into a geographical information system (GIS), one of the factors that must be considered is the state-of-the-art in computer technology. Operations that once took hours or days on mainframes can now be performed in real time or near-real time on microcomputers and workstations (43). In addition, many operations that were once considered impossible because of the limits of computing systems can now be revisited and, perhaps, implemented. The paper of Faust *et al.* (1991) attempts to describe the essential functionality required for remote sensing and GIS integration, evaluate the state of the art in computing technology, articulate current impediments to remote sensing and GIS integration, and define research issues that should be attacked to further this integration.

With the development of integrated sensor technology, satellite photogrammetry will gradually develop from the traditional ground control point control to orbit, attitude, and time-synchronous real-time control measurement and make satellite mapping gradually rid of the ground control point; the future intelligent remote sensing satellite will support remote sensing satellite from observation to decision "fast", "precision", "flexible" intelligent application service and realize global remote sensing data from access to application users of remote sensing information service in minutes and real-time mapping service.

Saunders and Seguin (1992) briefly outlined some promising future developments in the use of satellite data in meteorology and climatology during the next 10 years. Real-time information regarding the Earth's surface can be obtained periodically from satellite remote sensing. GIS can automatically provide many maps of agriculture, forest and hydrology, and land use information. Combining the experts' knowledge ("expert system"), further crop estimation and flood forecasting, etc., can be carried on by the integration of GIS and remote sensing. Although real-time imaging and mapping have been widely used for real-time flood forecasting and hazard mapping, crop diseases and insect pests surveillance, and forest fire monitoring, the integration of remote sensing, GIS, and "expert system" techniques for such purposes is an active area of current research (see Kimes *et al.*, 1991; Photogrammetric Engineering and Remote Sensing, 1991).

References

Babey, S. K. and Anger, C. D. (1989). A compact airborne spectrographic imager (CASI). In *Proceedings of the IGARSS'89 12th Canadian Symposium on Remote Sensing*, Vancouver, B.C., pp. 1028–1031.

Baudin, G., Bessudo, R., Cutter, M. A., Lobb, D., and Bezy, J. L. (1990). The medium resolution imaging spectrometer (MERIS). In *Proceedings of a Remote Sensing Society Workshop Held at Millband Tower*, London, pp. 32–46.

Borkowski, K. M. (1989). Accurate Algorithms to Transform Geocentric to Geodetic Coordinates. *Turon Radio Astronomy Observatory; Bull Geoid*, Vol. 63, pp. 50–56.

Borstad, G. A., Hill, D. A., and Kerr, R. C. (1989). Use of the compact airborne spectrographic imager (CASI): Laboratory examples. In *Proceedings of the IGARSS'89 12th Canadian Symposium on Remote Sensing*, Vancouver, B.C., pp. 2081–2084.

Brivio, P. A., Dellaventura, A., Rampini, A., and Schettini, R. (1992). Automatic selection of control-points from shadow structures. *International Journal of Remote Sensing,* **13**: 1853–1860.

Cracknell, A. P. (1988). Remotely sensed data for GIS. In Vaughan, R. A., and Kirby, R. P. (Eds.), *Geographical Information System and Remote Sensing for Local Resource Planning*, pp. 35–43. Dundee: Remote Sensing Products & Publications Ltd.

Dall, J., Jorgensen, J. H., Christensen, E. L., and Madsen, S. N. (1992). Real-time processor for the Danish airborne SAR. *IEE Proceedings-F Radar and Signal Processing*, **139**: 115–121.

Dozier, J. and Goetz, A. F. H. (1989). HIRIS-Eos instrument with high spectral and spatial resolution. *Photogrammetria*, **43**: 167–180.

Faust, N. L., Anderson, W. H., and Star, J. L. (1991). Geographic information-systems and remote-sensing future computing environment. *Photogrammetric Engineering and Remote Sensing*, **57**: 655–668.

French, M. N. and Krajewski, W. F. (1994). A model for real-time quantitative rainfall forecasting using remote-sensing. 1. Formulation. *Water Resources Research*, **30**: 1075–1083.

French, M. N., Andrieu, H., and Krajewski, W. (1994). A model for real-time quantitative rainfall forecasting using remote-sensing. 2. Case-studies. *Water Resources Research*, **30**: 1085–1097.

Gitelson, A., Garbuzov, G., Szilagyi, F., Mittenzwey, K. H., Karnieli, A., and Kaiser, A. (1993). Quantitative remote-sensing methods for real-time monitoring of inland waters quality. *International Journal of Remote Sensing*, **14**: 1269–1295.

Hamill, T. M., Dentremont, R. P., and Bunting, J. T. (1992). A description of the air-force real-time Nephanalysis model. *Weather and Forecasting*, **7**: 288–306.

Hielkema, J. U. and Snijders, F. L. (1994). Operational use of environmental satellite remote-sensing and satellite-communications technology for global food security and locust control by FAO — The ARTEMIS and DIANA systems. *ACTA Astronautica*, **32**: 603–616.

Hou, Y. T., Campana, K. A., Mitchell, K. E., Yang, S. K., and Stowe, L. L. (1993). Comparison of an experimental NOAA AVHRR cloud dataset with other observed and forecast cloud datasets. *Journal of Atmospheric and Oceanic Technology*, **10**: 833–849.

Inoue, Y., Sakuratani, T., Shibayama, M., and Morinaga, S. (1994). Remote and real-time sensing of canopy transpiration and conductance — Comparison of remote and stem-flow gauge methods in soybean canopies as affected by soil-water status. *Japanese Journal of Crop Science*, **63**: 664–670.

Gower, J. F. R., Borstad, G. A., and Hollinger, A. B. (1985). Development of an imaging optical spectrometer for ocean and land remote sensing. In *Proceedings of the 3rd International Colloquium on Spectral Signatures of Objects in Remote Sensing*, Les Arcs, France (ESA SP-247), pp. 219–225.

Joerg Kilngenfuss Pub. (1987). *Klingenfuss Guide to Utility Stations*. Joerg Kilngenfuss Pub.: Tuebingen, Germany.

Kimes, D. S., Harrison, P. R., and Ratcliffe, P. A. (1991). A knowledge-based expert system for inferring vegetation characteristics. *International Journal of Remote Sensing*, **12**: 1987–2020.

Klaes, K. D. and Georg, R. (1992). An efficient algorithm to process NOAA-AVHRR data in real-time. *Remote Sensing of Environment*, **39**: 75–80.

Kunkel, B., Blechinger, F., and Viehmann, D. (1991). ROSIS imaging spectrometer and its potential for ocean parameter measurements. *International Journal of Remote Sensing*, **12**: 753–761.

Kraus, K. (1992). *Photogrammetry — Volume 1: Fundamentals and Standard Processes, Bonn*. Germany: FERD. DÜMMLER Verlag.

Ladouceur, G., Allard, R., and Ghosh, S. (1986). Semi-automatic survey of crop damage using color infrared photography. *Photogrammetric Engineering & Remote Sensing*, **52**: 111–115.

Laxon, S. (1994). Sea-ice altimeter processing scheme at the EODC. *International Journal of Remote Sensing*, **15**: 915–924.

Liljas, E. (1987). Multispectral classification of cloud, fog and haze. In Vaughan, R. A. (Ed.), *Remote Sensing Applications in Meteorology and Climatology*, pp. 301–319. Hingham MA: Kluwer Academic.

Mashyanov, N. R. and Reshetov, V. V. (1995). Geochemical ecological monitoring using the remote-sensing technique. *Science of the Total Environment*, **159**: 169–175.

McGinley, J. A., Albers, S. C., and Stamus, P. A. (1991). Validation of a composite convective index as defined by a real-time local analysis system. *Weather and Forecasting*, **6**: 337–356.

NERC Satellite Station, Dundee University. (1993). Network access to NOAA HRPT data. NERC Satellite Station, Dundee University, Dundee, UK.

Pearson, R., Grace, J., and May, G. (1994). Real-time airborne agricultural monitoring. *Remote Sensing of Environment*, **49**: 304–310.

Photogrammetric Engineering and Remote Sensing. (1991). *Special Issue: Integration of Remote Sensing and GIS*, **57**: No. 6.

Ralston, A. (1965). *A First Course in Numerical Analysis*. New York: McGraw-Hill Book Company.

Ropelewski, C. F. (1991). Real-time climate monitoring of global snow cover. *Global and Planetary Change*, **90**: 225–229.

Sabins, F. F. (1987). *Remote Sensing: Principles and Interpretation.* New York: W. H. Freeman.

Salomonson, V. V., Barnes, W. L, Maymon, P. W., Montgometry, H. E., and Ostrow, H. (1989). MODIS: Advanced facility instrument for studies of the Earth as a system. *IEEE Transactions on Geoscience and Remote Sensing*, **27**: 753–761.

Saunders, R. W. and Seguin, B. (1992). Meteorology and climatology. *International Journal of Remote Sensing*, **13**: 1231–1259.

Schowengerdt, R. A. (1983). *Techniques for image Processing and Classification in Remote Sensing.* Orlando: Academic Press.

Shahidi, F., Woolliscroft, L. J. C. and Stadsnes, J. (1991). Interactive imaging and real-time pointing in an auroral imaging observatory. *ESA Journal-European Space Agency*, **15**: 141–148.

Sherwin, T. J., da Silva, J. C. and Robinson, I. S. (1995). The use of a fast-delivery ERS-1 SAR image to study internal waves at sea. *Earth Observation Quarterly*, **48**: 21–24.

Simpson, J. J. and Harkins, D. N. (1993). The SSABLE system — Automated archive, catalog, browse and distribution of satellite data in near-real time. *IEEE Transactions on Geoscience and Remote Sensing*, **31**: 515–525.

Singal, S. P. (1991). SODAR studies of sea breeze circulation on Indian coastline. *Indian Journal of Radio & Space Physics*, **20**: 397–401.

Sinha, P., Gorinsky, S. V., Laplante, P. A., Stoyenko, A. D., and Marlowe, T. J. (1996). Classification and overview of research in real-time imaging. *Journal of Electronic Imaging*, **5**(4): 466–478.

Smith, W. L. (1986). Evaluation of the effects of photo measurement error on predictions of stand volume from aerial photography. *Photogrammetric Engineering & Remote Sensing*, **52**: 401–410.

Soukeras, S. and Hayes, L. (1991). On-line access to the records of the University of Dundee AVHRR archive. *International Journal of Remote Sensing*, **12**: 183–191.

Sun, Xiuhong, and Anderson, J. M. (1993). A spatially variable light-frequency-selective component-based, airborne pushbroom imaging spectrometer for the water environment. *Photogrammetric Engineering & Remote Sensing*, **59**: 399–406.

Vane, G. (Ed.). (1987). The airborne Visible/Infrared Imaging Spectrometer (AVIRIS): A description of the sensor, ground data processing facility, laboratory calibration, and the first results. JPL Publication.

Valks, P. J. M., Piters, A. J. M., Lambert, J. C., Zehner, C., and Kelder, H. (2003). A fast delivery system for the retrieval of near real-time ozone columns from GOME data. *International Journal of Remote Sensing*, **24**: 423–436.

Whitehouse, B. G. and Landers, P. M. (1994). Transferring Earth observation imagery to the coastal zone via cellular telephony. *Marine Technology Society Journal*, **28**: 78–81.

Xue, Yong. (1995). Thermal modelling of the Earth's surface, Earth surface temperature determination and dynamic aspects study of surface temperature from remotely-sensed data. PhD thesis, University of Dundee, Dundee, UK.

Zibordi, G. and Meloni, G. P. (1991). Classification of Antarctic surfaces using AVHRR Data — A multispectral approach. *Antarctic Science*, **3**: 333–338.

Chapter 2

Integration of Remote Sensing, Geographic Information Systems, and Global Navigation Satellite Systems

2.1 Introduction

Increasing interest is being shown in the combined use of remotely sensed images and Geographic Information System (GIS) based dataset for environmental applications. The number of recent publications with an "integrated" label is increasing rapidly and shows the many benefits to be gained from integration. The functionality required in a GIS for environmental applications is somewhat different from that needed in other disciplines, such as market analysis, social geography, or urban planning, although there is some overlap. In particular, there is much more need for raster processing facilities because environmental parameters generally are continuous and, in natural/semi-natural habitats particularly, have no firmly located boundaries. However, distinct political or administrative boundaries are often imposed for the monitoring and/or control of natural areas-Bodies responsible for such natural areas, for example, the Nature Conservancy Councils, the National Rivers Authority, or the National Trust will have a requirement to monitor the area within their boundaries and will also look at the influence of surrounding areas (for example, the impact of farming within a buffer zone outside sites of special scientific interest). Thus, there is a clear requirement for the use of data in raster, vector, and tabular formats (Hinton, 1996).

A GIS is defined as a system that analyzes and displays spatially referenced data (Burrough, 1986). Remotely sensed images are forms of spatially referenced geographical information but most current GISs do not contain sufficient image processing capabilities for their analysis. Although there is probably no universally accepted definition of an Integrated GIS, Ehler *et al.* (1989) in a much-cited paper defined three levels of integration:

Level l: Separate database, cartographic, and image processing systems with facilities to transfer data between them. This was the usual situation until GIS systems with combined cartographic display and database management appeared.

Level 2: Two software modules (GIS and Image Processing) with a common user interface and simultaneous display.

Level 3: The long-term goal of full integration which would be achieved with a single software unit with combined processing.

Abel *et al.* (1994) stated that systems' integration involves the coupling of the pre-existing system. Similarly, Trotter (1991) claimed that integration is defined as the inter-change of data between respective systems but also states that, at its most advanced, full integration will provide an environment largely free of the logistical considerations of moving data between the two systems. There is, therefore, still some confusion over what does and does not constitute "integration".

2.1.1 *Historical route towards integrated GIS*

The term integrated GIS (or IGIS) has been used for some time (e.g., Faust *et al.*, 1991; Lauer *et al.*, 1991; Star *et al.*, 1991) inside the remote sensing community to describe either a link between an image analysis system and a GIS or a more "fully integrated" system. There is, of course, no reason why it should have been appropriated in this way. The use of the term IGIS by users of remote sensing "...indicates that they consider this particular integration to be special, in comparison to the myriad other types of integration accomplished, ongoing, or anticipated in GIS" (Dobson, 1993). Many authors are now referring to integrated GIS and expert systems (Hartnett *et al.*, 1994), GIS and mathematical models (Merchant, 1994), or GIS and spatial data analysis (Goodchild *et al.*, 1992; Densham, 1994).

Historically, GIS and remote sensing are closely linked. Early work leading to the development of GIS revolved around methods to better access aerial photographic coverage of specific areas (Star *et al.*, 1991). However most, although not all, geographic information systems available are based on a vector data model. While many now incorporate the facilities to store and display raster data, they do not include image processing functionality. Data stored in a GIS need to be converted into raster format in order to be integrated into the image processing chain. Integration of remote sensing information into a GIS occurs naturally in a raster GIS because the two data structures are similar. Integration into a vector system requires somewhat more effort but has been achieved by some vendors at least to the extent of updating vector information by using an image as a backdrop for vector editing and on-screen digitizing (Faust *et al.*, 1991). Some systems at "Level 2" Ehlers *et al.* (1989) are also emerging. These tend to be led from the image processing side with the incorporation of vector display facilities in raster image analysis systems and some use of the vectors in image processing.

More than 20 years ago, it was suggested that classification could be improved through the inclusion of ancillary data as another image band (Hutchinson, 1982). That work was performed entirely in raster with image layers, derived from ancillary data, in logical combinations to alter the results of the classification if necessary and correct some misassignments between classes. In 1987, a conceptual framework for a GIS was described as a connected image analysis system and GIS with data transfer between them (Archibald, 1987). However, the same author emphasized that the objective of a useful interface is not simply to pass manipulated remote sensing data from an image analysis system to a GIS, while limiting the use of data from the GIS for activities such as image registration. Integration will optimize the extraction of information from imagery by utilizing all of the related data and functionality that the GIS can provide. The trend to more emphasis on the application of GIS systems in part from improvements in the spatial and spectral resolution and the coverage of remotely sensed data available together with improvements in hardware and software to handle them. There is general recognition of the global nature of many environmental problems and hence larger and larger databases are being connected to provide information at various scales (Archibald, 1987). Green (1992) has suggested that effective integration of image processing and GIS tools can decrease the cost of gathering resource information, reduce the time required to capture the information, and increase the detail of the information.

2.1.2 *The need for integrated GIS*

There has been a rapid increase in the number of published papers describing the results of analysis using integrated remote sensing and GIS. However, many report restrictions are caused by a lack of closer integration of the data formats and the need for transfer of data between systems. Those carried out on "level 2" type systems report the requirements for format conversions and costly overlays, but with dual format storage and display capabilities, these systems do provide a temporal solution. The processing and analysis of data within dual format systems generally take place in either vector or raster mode. Rarely do these systems provide the functionality to perform a procedure using both vector and raster data simultaneously (Piwowar and LeDrew, 1990).

Many scientists already use a number of different image analysis systems and GISs in a single project to take maximum advantage of available functionality. The recent trend has been towards multisystem use of multiformat spatial databases. In practice, sharing spatial data is difficult because of the unique way each system handles storage and processing. The ability to transparently access foreign data is one way forward. However, there are some potential problems that were identified by Lauer *et al.* (1991).

(a) There can be differences in the spatial model even within the same generic format (e.g., object-oriented (Worboys *et al.*, 1990) versus geo-relational (Burrough, 1986) vector models) which can be real barriers to information exchange at a user's level. While it is being made more straightforward by system manufacturers to transfer vector lines, points, and areas between systems, it can still be a problem to transfer the attribute database along with the geometries.
(b) Conversion of data between formats may often lead to some generalization and loss of accuracy. To minimize any loss of quality, the data should be left in their native format.
(c) Systems are still either vector-based or raster-based and this tends to limit the applications to which the system is put. This is the case even in "Level 2" image analysis systems which can access and display

vector data from a GIS system but have limited vector GIS functionality for spatial decision analysis. Abel *et al.* (1994) also admit that current GIS has a basic weakness as components of integrated systems, requiring fundamental extensions to achieve effective integration.

In "an example of integration of remote sensing and GIS technologies" for habitat studies, Gagliuso (1991) described substantial manual and digital cartographic Interaction in the studies together with data transfer between formats and systems because the required true integration of the two technologies was not available. Similarly, Baumgartner and Apfl (1994) created snow maps by classifying satellite image and then vectorized them and transferred them "via interfaces" to the GIS module. This sequence is common in many papers on "integration" and has its benefits. However, the processing steps must be well defined in advance and are thus often inflexible if the user discovers the need to take several steps back to the image processing system. Chagarlamudi and Plunkett (1993) have emphasized the need for low-cost mapping solutions for developing countries. This usually requires PC-based solutions. They report that a principal problem in effectively integrating remotely sensed and map data in digital form on the PC systems available has been in the exchange of data between the raster-based image processing system and the vector-based GIS.

Remote sensing analysis is increasingly using GIS data. Examples of this include the over of vector data for presentation and location and the use of digital elevation models (DEMs) for terrain correction of imagery and perspective visualization. However, there is still considerable uncertainty over how data held in these different models should be analyzed together. It is a commonly held belief that with remote sensing the main processing tasks are concerned with the labeling of each pixel but this is not necessarily so. For the increasingly popular and widely available data from radar sensors, it is more meaningful to look at real characteristics since the speckle inherent in such data makes it difficult to analyze on a per-pixel basis. It is usual to extract backscatter and texture information from the image for comparison with other characteristics of the area. Similarly, remotely sensed images are being used to extract landscape characteristics such as road locations or field boundary information to vector line maps (see Section 2.4). There is no longer a need to work just in pixel mode with remotely sensed data.

As an example of this, Jansen *et al.* (1990) used topographic map data to classify an image on a per-polygon basis. However, as the systems they used have no capabilities for integrated processing of this type, they rasterized map objects and then used these as a mask within a self-developed program for assessing the dominant class within objects from an image classified using a standard per-pixel classification process. Post-classification altering was necessary to assign the most common pixel value as the final class of the polygon. The authors admit that this is far from an optimal route but the necessary systems were not available. In a similar analysis, Kontoes *et al.* (1993) used GIS-derived data to post-process the classified image. Data from both a raster-based image processing system and a vector GIS were combined in a "reasoning program". Ancillary data sources originating from maps "...had to be digitized, pre-processed, and converted into co-registered raster form in order to be integrated into the classification methodology". By using data interchange format, polygon boundaries are stored in a GIS, and in some cases, the attributes attached to those polygons can be transferred to the image analysis system. There, this information may be used in, for example, image segmentation, to aid the selection of training sites in supervised classification or for some form of image enhancement (Trotter, 1991). However, once transferred to the image analysis system, GIS functions can no longer be applied to them (e.g., selection/editing, attribute query, topological queries, and buffering). Despite the vast amount of digital ancillary information often available and its potential usefulness in image classification, applications which use these data are rare and often simplistic (Trotter, 1991). Improvements in the ease of integration through the development of integrated systems should help change this situation.

Mattikali (1994) describes the use of an integrated GIS. However, the author states that "...data derived from remotely sensed satellite images ...are integrated into the vector GIS" and that "...the integration required an interface to be developed between the vector GIS and the raster image processing system". It is questionable whether this should be referred to as an integrated GIS (Hinton, 1996). Johansen *et al.* (1994) have presented GIS as a tool for the integrated analysis and interpretation of remote sensing-based maps with georeferenced *in situ* or model environmental information. In their analysis which involved unsupervised and supervised classifications and image filtering, the authors used a vector-based GIS, a raster-based image processing system, and a data visualization program. The systems were "integrated" by the passing of data between them. This approach to integration has disadvantaged but is necessary until fully integrated systems with capabilities for both raster and vector analysis become the norm.

2.1.3 *The complementarity between GIS and remote sensing*

As the demand for spatial information grows, there is an ever-increasing synergy between remote sensing and geographical information systems. There are three main ways in which remote sensing and GIS technologies are complementary to each other:

(a) remote sensing can be used as a tool to gather datasets for use in GIS,
(b) GIS datasets can be used as ancillary information to improve products derived from remote sensing data, and
(c) GIS data can be used together in environmental modeling and analysis.

There is a close interrelationship between these three types of complementarity. There have recently been a number of important developments in these areas. It has been suggested that the synergism between these two technologies is the major advantage of continuing to improve integration (Star *et al.*, 1991; Trotter, 1991). Some examples of integrated procedures are as follows:

(a) Line extraction from images to form boundaries.
(b) Per-polygon classification.
(c) Use of DEM's to correct image illumination/aspect or to geometrically correct radar images to allow for radar imaging geometry.
(d) Use of image, DEM, or shading information as the basis of vector search and query, such as selection of polygons above a given mean height derived from the DEM. If a system is fully integrated, it should be possible to use the information in its native format and perform such queries "on the fly".
(e) Modeling studies, e.g., editing height attributes of vector areas after erosion scenarios, have changed the DEM.

These are discussed in detail in the following sections.

2.2 Remotely Sensed Data as an Information Source to GIS

The potential for using digital remotely sensed data as a primary information source for GIS has received considerable attention in recent years, particularly in applications related to natural resources and at large spatial scales. Data are still the most expensive component of a GIS, and the need for inexpensive, accurate, and timely GIS data has created a collateral demand for digital images (Green, 1992). However, the majority of GIS continue to use a vector data structure and are thus separated from remotely sensed data at this fundamental level. This was suggested as an impediment to integration by Ehlers *et al.* (1989); Trotter (1991), more optimistically, suggested that further developments in both research and technology will be required before fully operational procedures for the automated acquisition of GIS information from image data can be established but that these developments are foreseeable. The difference in scales commonly encountered in GIS and remote sensing also needs investigation.

2.2.1 *Thematic mapping*

Thematic mapping from space-based sensor systems is one of the main use of remote sensing. The datasets so produced can be used as additional layers of data in a GIS. However, the techniques for generating thematic maps from satellite data still suffer from several problems, and a number of studies have pointed to the need for further research and development (e.g., Trotter, 1991). The core components of thematic mapping from satellite-generated data are, apart from data preprocessing operations, (i) classification and (ii) spatial generalization. There have been very recent and important developments in both of these areas.

2.2.1.1 *Recent advances in classification*

The classification of remotely sensed data has been the subject of much investigation over the past two decades since the launch of the first Landsat in 1971. Regrettably, although many important developments have taken place, especially in the last five years, most commercial software packages for the processing of remotely sensed data still rely on the early classification methods based on the digital count values of the pixels in the different spectral bands of the sensor in question. Typically, the distribution of the image pixels is modeled in spectral feature space using parametric classifiers such as the minimum Euclidean distance classifier or the maximum-likelihood classifier. The most important and potentially valuable departures from these methods are (not in any order of importance) as follows: (a) the use of spatial or textual information from the image space (including fractal measures), (b) the use of knowledge-based and evidential reasoning methods, and (c) the use of artificial neural networks.

There are many ways to compute texture features from the neighborhood of image pixels to aid in classification and a common approach is to use the statistical measures computable from the gray-level co-occurrence matrices (Haralick *et al.*, 1973; Franklin and Peddle, 1990). Such texture measures can be routinely calculated and utilized alongside spectral data in classification. Normally, the use of such measures yields improvements in total classification accuracy. While such measures are relatively easy to calculate, they are not known to correlate as well with human perception of textures as other methods based on fractals (Pentland, 1984, 1985). Since it is now well-recognized that fractal patterns occur widely in nature, it is appropriate that they should be considered a suitable tool for the analysis of spatial patterns in the landscape viewed from space (Wilkinson *et al.*, 1995). Several studies have recently shown that fractal texture measures can be used successfully as additional image features to improve classification (e.g., De Cola, 1989; Priebe *et al.*, 1993). Recent work has also shown the value of "multifractal" rather than "monofractal" models, e.g., for analysis of textures in Synthetic Aperture Radar (SAR) images (Arduini *et al.*, 1991).

Knowledge-based and evidential reasoning methods for image classification have been tried and evaluated in a number of experiments over the past decade. Knowledge-based methods concern the use of ancillary data, heuristics, and facts, which may be very simple or highly complex. Evidential reasoning is an approach that has become relatively popular for combining multiple pieces of evidence coming from background knowledge and ancillary information. Early work in the application of knowledge-based methods to remote sensing was related to the interpretation not of satellite imagery but of aerial imagery and was mostly concerned with recognizing man-made features (e.g., roads) using knowledge or models about expected shape or structure. Typical of this genre is the work of Bajcsy and Tavakoli (1976), Fischler *et al.* (1981), Harlow *et al.* (1986), Perkins *et al.* (1986), Nicolin and Gabler (1987), and Matsuyama (1987) among others. Such approaches were also applied to the interpretation of satellite imagery (e.g., McKeown, 1987; Goodenough *et al.*, 1987; Wang and Newkirk, 1988; Wu *et al.*, 1988; Van Cleynenbreugel *et al.*, 1990).

Recently knowledge-based methods have been applied to the analysis of spectral profiles from airborne imaging spectrometers for classification purposes (e.g., Chiou, 1985; Kimes *et al.*, 1991, 1992). Such an approach will also gain in importance because of the trend towards higher numbers of spectral bands in future satellite sensors.

The use of artificial neural networks has been another important and relatively recent advance in the development of classification methods for thematic mapping from satellite image data. Neural networks have been found to perform well as general pattern recognition systems in many disciplines and there were significant developments in the 1980s (Dayhoff, 1990; Pao, 1989). There are many different types of neural networks which are appropriate for different types of problems. They all, however, consist of interconnected networks of simple processing elements which can be used to classify or cluster datasets. The most commonly used type for satellite image analysis is the multilayer perceptron model, also called the semi-linear feed forward network, based on several layers of interconnected processing elements which carry "weight" attached to their connections. Once trained, these weights can be used to encode a mapping from a pattern space to a classification space. One of the most useful breakthroughs in the development of these networks for pattern recognition was the invention of the backpropagation training algorithm (e.g., Rumelhart *et al.*, 1986).

Neural networks began to be used for the classification of remotely sensed satellite data in the late 1980's and a growing body of literature has been steadily accumulating ever since demonstrating their potential (e.g., Key *et al.*, 1989; Lee *et al.*, 1990; Hepner *et al.*, 1990; Benediktsson *et al.*, 1990; Bisehof *et al.*, 1992; Kanellopoulos *et al.*, 1992; Downey *et al.*, 1992; Civco, 1993; Benediktsson *et al.*, 1993; Yoshida and Omatu, 1994; Foody, 1995; Chen *et al.*, 1995). In most reported cases, the neural network classifier has achieved comparable total classification accuracies to conventional parametric classifiers, with, in some cases, very significant improvements. Even though the division of feature space made by a neural network is unlikely to be optimal (this could require infinite training time), the results achieved are usually found to be of high quality, although care has to be taken in using them to avoid overfitting training data so that the classifier will not generalize well to new examples. It has also been found in some cases that conventional parametric classifiers such as the maximum-likelihood classifier can have complementary potential to the neural network and it has been found possible, and beneficial, to integrate them (Kanellopoulos *et al.*, 1993; Wilkinson *et al.*, 1995a). Indeed, in view of the very different ways in which neural networks and statistical classifiers divide feature space, this integrated approach now seems to be optimal.

Neural networks also have two important characteristics which, apart from their generally good performance at classification, may make them important tools in the future analysis of remote sensing data for thematic mapping. The first is that they do not depend on any statistical model of the dataset concerned. This means they can be particularly suitable for mixed datasets comprising multisensor image data such as combined Landsat Thematic Mapper (TM) data and ERS-l Synthetic Aperture Radar (SAR) data for which the underlying statistical property is not well known. The use of neural network classifiers in such a context has already been demonstrated (Wilkinson *et al.*, 1993, 1995b). The second important characteristic of neural networks is that they are suitable — in principle — for exploiting parallel computer hardware. Even though more research is still needed on developing good methods for training neural networks in parallel, the possibility of easily using parallel hardware will be important in the future as the volume and complexity of remote sensing and GIS datasets grow. It should be noted that there is a need for new software tools to enable neural networks to be used easily by practitioners in remote sensing and GIS.

2.2.1.2 *Recent advances in spatial generalization*

Since the sensor systems onboard satellites digitally sample the scene viewed by the optical systems, the images produced and relayed to the ground have a raster structure based on pixels. This data structure is of course very different from the usual vector data structure of a digital map. While in some geographic areas the landscape may consist of large homogeneous parcels, in many places, typically in southern Europe, the landscape can be so broken or inhomogeneous that thematic mapping must be carried out initially from a per-pixel classification of the scene.

The generalization of the resulting pixel-based thematic map is a difficult problem. Generalization is needed essentially to transform the map into a form which is suitable for vectorization and storage in a GIS. Even though some GIS are raster-based or support hybrid data structures such as quad-trees, vectorization is still a commonly applied procedure. Generalization in the image domain can therefore be viewed as the process of preparing the thematic map data for entry into the GIS. This is needed essentially to do two things: (i) to remove spatial noise arising from erroneous classifications and (ii) to reduce the overall images' entropy in order that it can be converted into vector form which (a) needs a much lower storage requirement than in raster form and (b) is readily understood by the end uscr in conventional cartographic terms. This generalization process is extremely difficult to carry out.

Various techniques have been proposed in the last 10 years to improve the spatial quality of satellite-derived thematic maps. These techniques only apply at the low or local level and are concerned with creating regions with a high degree of homogeneity. They include spatial altering and segmentation. Spatial altering refers to the class of techniques concerned with applying simple numerical procedures in a neighborhood around a particular pixel in order to remove entropy or "disorder". At its simplest, spatial altering in the classification or thematic map domain involves the removal of "isolated" pixels. These are pixels surrounded by pixels of a different class. Where the surrounding pixels all belong to the same thematic map class, it is easy to reassign the central pixel to their class. If the surrounding pixels do not all belong to the same class, it is possible to reassign the central pixel to the majority class. This can even be done iteratively and is known as Iterative Majority Filtering (Goldberg *et al.*, 1975; Guo and Moore, 1991). It is surprisingly efficient at smoothing the overall structure of a thematic map. In some cases, this type of procedure has the undesirable side effect of reducing the area (or the number of pixels) assigned to certain classes. Techniques have therefore been proposed to deal with this problem by iteratively re-assigning pixels to the underpopulated classes but in such a way as to keep the spatial structure simple (Wilkinson, 1993a).

The second class of spatial operation in the raster image domain is segmentation, which is usually applied before classification. Segmentation has taxed the best minds of the image processing community for more than two decades and even now it cannot be performed totally reliably, but several approaches are now beginning to show much promise. Segmentation has traditionally followed the two independent approaches based on (a) the detection of edges or discontinuities in the images or (b) growing regions by starting from seed points and adding more and more adjacent pixels until a discontinuity is reached. These approaches are notoriously difficult to get right because of the following:

(i) Edge detection generally is very sensitive to the thresholds used to define an edge (too low and the image is covered with too many unmanageable edge pieces with no obvious structure; too high and it is left with a few disconnected edge fragments which are difficult to join into a meaningful structure by any automatic algorithm).
(ii) Region-growing is dependent on which seed points are selected and in which order the neighboring pixels are scanned to see if they should join the spatial group which is in formation.

The vector processing capabilities of GIS can be used to define areas in an image for the processing discussed above. However, use of pattern recognition, edge extraction, and segmentation algorithms must all be possible and the results thereof are used together with the images. Edwards and Beaulieu (1989), for example, described a method to improve the classification of forests' clear cuts by image segmentation. However, to segment other than simple images, integrated GIS will have to expand functionality.

Notable among recent developments are those which integrate edge detection and region growing (Pavlidis and Liow, 1990), those which avoid order dependence in region growing by using the technique of iteratively scanning the entire image and looking in each cycle for the most similar pair of pixels to join together- the so-called "best merge" approach (Tilton and Cox, 1983; Beaulieu and Goldberg, 1989; Tilton, 1989), and those which make use of knowledge-based approach driven by existing maps stored in a GIS (Corr *et al.*, 1989). Edge detection filters can be applied to define domain boundaries from suitable mages which can then be used to update existing vector datasets (e.g., Jazouli *et al.*, 1994; Ton *et al.*, 1989) or to constrain the examination of other image data. In a fully integrated system, vector polygons extracted from images using edge detection methods can be modified according to a set of rules based on the value of the image pixels within them (Hinton, 1995b). Image segmentation polygons derived from optical imagery could be very useful for stratifying radar data, which are difficult to digitally segment because of their noise characteristics. A problem with raster-derived lines and boundaries is that their stepped appearance generally reveals their raster origin. However, vector GIS functionality should be able to incorporate pixel interpolation algorithms that produce smooth boundaries.

In some cases, it can be beneficial to smooth satellite imagery before any classification or segmentation operations are carried out, especially where the raw data are suspected to be noisy. For this purpose, many different kinds of alterations exist which have been developed over many years by the image processing community. However, of all the various approaches for doing this, the ones of most interest and most satisfactory with remotely sensed satellite imagery appear to be those based on (i) edge-preserving smoothing (Nagao and Matsuyama, 1979) and (ii) gradient morphology (Schoenmakers and Stakenborg, 1992).

At a simple level, this could be demonstrated by extending a GIS to support image-based map revision which requires an integrated raster-image/vector-graphics display with vector editing facilities (e.g., Derenyi and Pollock, 1990). The integration of landcover information in raster form (such as the results of a pixel classification) with vector GIS functionality can be of benefit in ecological studies. For example, for landscape pattern analysis, the patch size, frequency, land use classes, and lengths of boundaries between contiguous cover types are all important variables used in explaining certain ecological characteristics of a landscape (Griffiths and Wooding, 1988). Systems which can perform both vector topological queries and raster/vector intersection queries will give ecological researchers a substantial advantage. To compare a habitat map created from an image classification with an original vector habitat map, May (1986) converted the original map to raster format and performed pixel by pixel comparison of the rasterized original and the classification output. A fully integrated system is able to query the raster pixels within vector areas and perform such analyzes without format conversions and overlays. However, the generalization of classified digital image data to the mapping scales commonly encountered in GIS is considered a fundamental problem to be resolved before digital image data relating to natural resources can be incorporated automatically into GIS (Ehlers *et al.*, 1989).

An example of closer integration of remotely sensed image processing and GIS for map updating has been described by Brown and Fletcher (1994). Here, statistics from multispectral images within areas defined in a vector dataset are examined and if these statistics suggest a change in land use from that in the vector attribute database, the database is interactively updated. Janssen *et al.* (1993) performed a combined map update and classification procedure using edge detection procedures to define field boundaries in the vector GIS, performing a pixel classification of the image, calculating the most frequent class assigned to pixels within each of the detected adds, and saving that class as an attribute of the field polygon. A Global Change Database has been established (Hastings and Di, 1994) based on satellite data and designed for input into a GIS-Progress can be made in understanding the global environment if such data are developed for GIS use and GIS can be enhanced to handle them.

2.2.2 *Continuing problems in thematic mapping*

2.2.2.1 *Problems in classification*

Although as we have seen in Section 2.2.1 there are now a number of new and very promising techniques for classifying satellite data to make thematic maps, it is still true to say that satellite imagery often does not yield the quality of product needed by users of GIS. What are the reasons for this?

First, it has to be recognized that current satellite sensors are still quite limited in their ability to distinguish many types of land covers mainly because they lack the spectral sensitivity necessary to separate different types of vegetation, etc. The most advanced operational satellite sensor system from a spectral point of view remains at present the Landsat-TM instrument with its seven bands spread between the visible and thermal infrared. However, with the launch failure of Landsat-6 and with Landsat-5 being already seven years beyond its design lifetime, even this source of information may soon cease. Also, it is worth noting that, setting aside the TM thermal channel which has lower spatial resolution, the six principal bands each with 30 m resolution contain a certain level of redundancy and it is often found that they can be transformed into only three channels, e.g., by the Principal Components transform or the Tasseled Cap transform (Crist and Cicone, 1984) with minimal loss of information. Although the use of multitemporal or seasonal sequence of images through the year can help significantly (Megier *et al.*, 1991), this does not offer a complete solution to the problem since some areas of the landscape may change their characteristics through the year through human or natural interventions (re-planting, forest clearance, construction, burning, flooding, etc.), and second, the reliance on multitemporal sequences limits the generation of thematic products to once a year only which for some applications can be a drawback.

The other main difficulty with thematic mapping from space is the effect of the spatial resolution of the sensor. There are many conflicting aspects which make it difficult to select the ideal resolution for a given mapping task. On the one hand, the resolution improves more detail can be seen and there can be little doubt that the substitution of Landsat Multi-Spectral Scanner (MSS) images having 90 m resolution with TM images having 30 m resolution has been beneficial for mapping applications. However, as resolution improves, there is the problem of increase in data volume. Also, pixels which are mixtures of different land cover features at one resolution do not necessarily become un-mixed at a higher resolution — they simply become mixtures of different things. Hence, the improvement of resolution is not always beneficial. The increase in spectral variability for TM compared to MSS has been well documented (Irons *et al.*, 1985).

It is quite clear at present that remote sensing lacks an appropriate information theoretic model to deal with the combined issues of spectral and spatial resolution needed for specific mapping applications. The only quantitative approach which has so far been applied to the measurement of landscape variability and the space scale at which it should be ideally sampled is the semi-variogram method (Curran, 1988). It is possible that there is also a role for fractal models in this context since they are concerned with the relationships between spatial variability at different scales, though no firm conclusions have yet emerged on this. As new sensors are flown in the coming year, more experience will be gained by the remote sensing and GIS communities of different spectral–spatial resolution combinations and it remains to be seen how effective the new kinds of data (e.g., from the medium-resolution imaging spectrometers such as MODIS and MERIS) will be for thematic mapping purposes. Also, the problem of integrating and comparing remote sensing data at different space scales is becoming a serious concern and has recently received some much-needed attention (Raffy, 1994).

A further and most important problem in the classification of satellite imagery for thematic mapping is the adequacy of the ground data used in training the classifiers. Ideally, the ground data should give good example areas of each class desired in the thematic map product so that the classifier, of whatever type, can be trained for good recognition. However, it is frequently found that it is impossible to get a wide enough range of ground data to describe all of the physical variations seen within one thematic map class in the area of investigation.

2.2.2.2 *Problems in spatial generalization*

To a large extent, the techniques discussed in Section 2.2.1 for spatial generalization concern only low-level generalization in the image domain — that is to simplify the spatial structure based on a guiding principle that the thematic map should be made as visually simple as possible but on the basis that only information from the image domain enters the process. The process of mapping, to produce a GIS dataset, is different from this since it requires that information from the mapping domain be used. This distinction is frequently not understood.

The use of information from the mapping domain implies that information about the desired map scale, the desired map classes, and the constraints set by the map users on the minimum includable spatial unit, etc. are used. It is this link between the image domain and the anal map domain that frequently causes serious problems in thematic mapping and results in maps generated by automatic analysis of satellite imagery not being used as GIS datasets.

2.3 GIS as a Tool for Remotely Sensed Data Analysis

The second way in which remote sensing and GIS technologies are complementary to each other is in the use of pre-existing GIS datasets in the interpretation of remotely sensed data. There are a number of areas where GIS data can be used to enhance standard image processing functions. These can be broadly separated into (i) geometric and radiometric correction, (ii) image classification, and (iii) "Area of interest" operations.

2.3.1 *Exploiting GIS data as ancillary knowledge*

As we have seen, the analysis of remotely sensed satellite imagery is a complex problem subject to many difficulties, especially as regards classification and generalization. However, pre-existing GIS data can have roles both in the performance of classification and also in spatial operations in the image domain.

2.3.1.1 *Using GIS data for image stratification*

The use of vector polygons to restrict the area of an image to be processed can be thought of as masking operations without raster masks. An example of this would be the application of a filter to an area of the image within a vector-defined area while leaving the image outside the selected area unaltered. In a poorly integrated system, this requires a raster mask to be generated for the area to be filtered, a band combination operation to extract the area of interest, altering the result, and a further combination to restore the unaltered area to the filtered image. In a fully integrated system, it should be possible to perform this task in a single step without the creation of additional data layers. The benefits of this level of integration are faster processing times, no requirements for disk storage for intermediate data formats, and the reduction of data integrity problems caused by having multiple copies of the same data in different formats.

The simplest way to use GIS data in the analysis of remote sensing data is in the stratification of the landscape and the imagery. This term refers to the technique of broadly dividing the landscape into major areas which have very different characteristics. These characteristics may be typically topographic, phenological geological, or climatological. The purpose of such stratification is to divide the image into zones which can be treated differently for the purposes of thematic mapping. These could be topographic zones such as lowland and upland districts, climatic zones, such as Atlantic, Mediterranean. For example, in mapping land cover, in southern France, from Landsat TM imagery, Hill and Mégier (1988) stratified an entire Département into 12 separate ago-physical zones which were treated separately in classification. These zones were mainly based on altitude since the test area spread from the broad, fertile, and low-lying valley of the River Rhone up to the mountainous plateau of the Central Massif. In this case, digital topographic information formed the basis of the stratification. In European scale projects, such as the European Community's MARS Project (Monitoring Agriculture by Remote sensing), climatological stratification based on data stored in a GIS forms an essential component in the analysis of the satellite data. In such examples, stratification can be used to adapt the classifier of the data to deal with the separate stratified zones in different ways in order to improve the total classification accuracy achievable.

2.3.1.2 *Boolean operations between data layers*

Related to stratification is the use of Boolean logic operation between data layers in deriving new spatial products. Boolean operations are common in the analysis of GIS datasets to derive new data layers. Likewise, they can be used to derive products from remotely sensed satellite imagery if the imagery is treated just like one of the data layers from the GIS. The Boolean operations are usually performed on a regular grid basis using either raster or vector data structures. The technique is also known as sieve mapping and has been applied successfully in many applications. For example, the technique has been used to predict areas of potential landslides in China on loess deposits based on combined data layers giving slope, aspect, and lithology (Shu-Quiang and Unwin, 1992). In this example, however, the model did not specifically include satellite data. In the work by Hill and Méger (1988), a set of six Boolean operations were defined to modify the land cover classes output from the image classifier. These operations directly incorporated the ago-physical zones produced from a preliminary stratification from topographic data.

2.3.1.3 *Image classification*

Perhaps the most obvious benefit of combining vector information in image classification is in the selection of training areas. To be of maximum benefit, the system used must be able to perform what has been called a raster/vector intersection query (Ehlers *et al.*, 1991) (i.e., given an image and a polygon file, which pixels fall within which polygon), without carrying out any data format conversions or raster overlays/combinations. For deciding which areas are suitable for use in image classification, whether the areas are characteristic of the class and whether they cover the full range of characteristics of the class, a method is required for examining image statistics within training polygons. These could be examined in conjunction with other attributes of the training area to make maximum use of other available pertinent information. Very few system developers have yet appreciated this need (although see Derenyi *et al.*, 1992; Hinton, 1995a). The ability to store statistics attributes "can aid qualitative decision-making and can facilitate change detection without performing the classification of multitemporal data sets" (Derenyi *et al.*, 1992). In an integrated system, it is possible to look at the change in image statistics within defined polygons directly without having to actually classify the image and examine the raster result.

Integration at Level 2 (Ehlers *et al.*, 1989) will allow vector data to be rasterized as "image planes" and incorporated into traditional classification and segmentation procedures. Conradsen and Gunulf (1986) improved the classification of remotely sensed data using additional variables. Kushwaha *et al.* (1994) incorporated image textural information as additional bands in an image classification. This approach has often improved classification accuracies. However, in a fully integrated system, it should be possible to use all GIS data in their native format, together with the image statistics derived for each polygon, in a classification of the polygon attribute table. This is illustrated schematically in Figure 2.1.

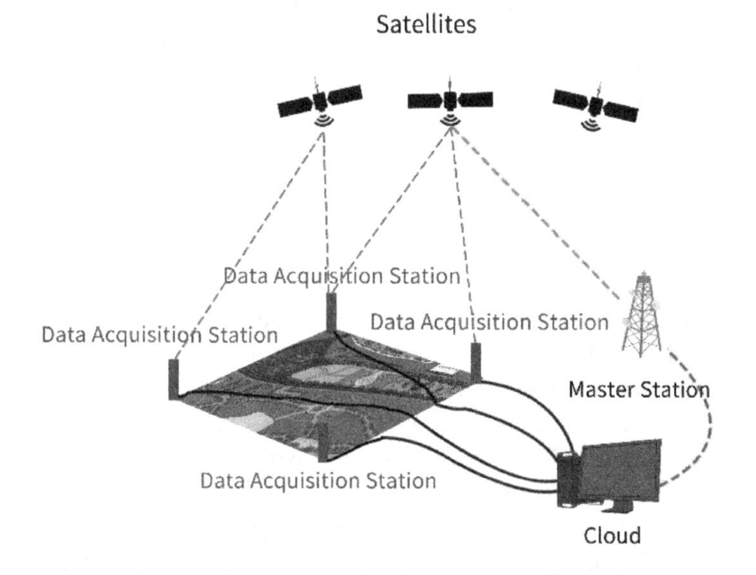

Figure 2.1. Schematic diagram of CORS positioning.

An important criterion of a fully integrated system is that the data flow must be two-way between the raster image and the vector dataset. It must be a simple operation to generate image statistics within polygons and then return those statistics directly to the GIS database as attributes of the polygons. For image classification, it must also be possible to exclude edge pixels from the calculation of these statistics in order to remove classification errors associated with mixed pixels at the polygon boundaries. A true polygon classification as illustrated in Figure 2.1 would be independent of the resolution of the images used. Statistics from a Thematic Mapperimage (30 m resolution) and from a SPOT Panchromatic image (10 m resolution) can be generated and stored as polygon attributes for use in classification, without the need for resampling the images to the same grid. It should also be possible to return classification results and accuracy statistics directly to the database as illustrated in Figure 2.l and as suggested by Archibald (1987) who stated that "in poorly integrated systems, classification results will have to be overlaid with map data so that they can be pointed back to polygons". It is important that the spatial distribution and magnitude of classification accuracy are known by users of the results of classification (Foody, 1988). Saving such results to the vector attribute database ensures they are easily accessible.

Since neural network classifiers behave as general pattern recognition systems and, moreover, they assume no prior statistical model for their input data, they are extremely flexible when it comes to mixing different kinds of data. For this reason, they can be used to integrate GIS data in remotely sensed image classification (Wilkinson, 1993b). A few experiments have already been reported on this use of GIS data. For example, Benediktsson *et al.* (1990) used topographic data alongside Landsat Multi-Spectral Scanner (MSS) data in ground cover mapping. In another experiment (JRC, 1991), terrain height from a digital terrain model (DTM) was used as an extra input to a multilayer perceptron neural network alongside two date SPOT HRV imagery. The objective of the experiment was to classify land cover in the satellite imagery. In this case, the inclusion of the DTM information from a GIS both increased the overall classification accuracy and furthermore reduced the training time needed for the neural network to half. So far the use of GIS data in neural networks is a relatively new development and very little has so far been attempted. However, the limited evidence available points to the fact that this would be a profitable approach to take.

2.3.1.4 *Using evidential reasoning with geographic context*

The use of sieve mapping or Boolean operations for combining GIS data layers and remotely sensed data is an entirely deterministic procedure which does not suit all applications. A much more flexible way of treating combined layers of data and ancillary facts or information is to use a knowledge-based or expert system approach incorporating a procedure for making complex inferences when the information available is incomplete or uncertain. As stated in Section 2.2.1, such methods have already been applied extensively in image analysis, e.g., to exploit ancillary information about the expected shapes of features in imagery. They can also be used to integrate GIS datasets in the image classification process. This is because the data in the GIS can provide context which can be utilized in refining image products, not just on the basis of simple stratifications or on Boolean logic but also on the basis of complex contextual rule bases. The reason for choosing this kind of approach is that much of the context information available regarding particular image objects (pixels, segments, etc.) may be only indicative of possibilities rather than offering definite yes/no answers.

Furthermore, when there is a large body of evidence available and some of it is contradictory, a problem-solving approach is definitely needed. Even worse, it is sometimes necessary (and this especially applies very often in the thematic mapping) to integrate evidence that applies at different hierarchical levels in a classification scheme. For example, evidence from a DTM may suggest that a particular region is too hilly to cultivate crops. On the other hand, evidence from a soil map may suggest that cereal crops in particular are favored here because of the texture and drainage conditions. If in addition an initial classification of the satellite data has yielded some pixels of wheat in this same area, there is a difficult problem to solve. The complete set of available evidence applies to different levels of the classification scheme (specific crop-class of crops-crops in general) and is contradictory across these levels. A promising and recently much-favored approach to adopt in such circumstances is the method of reasoning based on the Dempster–Shafer theory of evidence (Shafer, 1976). This method has several advantages over the more conventional Bayesian reasoning. In Dempster–Shafer evidential reasoning, evidence derived from facts in a knowledge base may be applied not only to singleton classes but also to groups of classes. The approach also handles uncertainty in an interesting and arguably highly realistic way. Lack of evidence for a particular class is not treated as evidence to be shared between all the other classes as done in the more classical approach. Instead, it is attached as evidence to the "frame of discernment" which is the super-class comprising all classes in the classification scheme. The Dempster–Shafer algorithm basically provides a mechanism for weighing up pieces of evidence coming from knowledge-based system rules, triggered, for example, by geographic context information from a GIS, and then deciding the optimum classification on the basis of computed "belief" values. The evidence may be both confirming or disconfirming of either specific classes or groups of classes. Also, if there is not enough information available to assign an individual class label to an image object, the method can assign a class group label and by so doing indicate that there is insufficient information available to be more specific.

The value of these techniques has been recognized in a number of recent studies involving knowledge-based interpretation of combined remote sensing and GIS data (Lee *et al.*, 1987; Garvey, 1987; Kim and Swain 1989; Moon, 1989; Moon, 1990; Srinivasan and Richards, 1990; Wilkinson and Meger, 1990; Peddle and Franklin, 1992; Loguercio-Polosa *et al.*, 1992; Kontoes *et al.*, 1993; Peddle, 1995, among others). In some recent work (e.g., Molenaar and Janssen, 1992), the value of using GIS "objects" to restructure remote sensing data as part of the interpretation process has been noted — this appears to be highly relevant to current needs, though developments are still needed in data models and representation to make this effective.

2.3.1.5 *Geometric and radiometric correction*

Remotely sensed imaging needs to be registered to some geographical coordinate frame to be of practical use. Some forms of remotely sensed data, and over some geographical areas, can be substantially affected by varying topography on the ground. It can be very difficult to apply remotely sensed data to land use and environmental mapping in mountainous areas because of these effects (Bertolo *et al.*, 1993) and yet it is in just these areas that the environmental information is difficult to obtain in practice by any other means. GIS data (vector, point, and area information. Topographic data are key attributes required for natural resource management (Trotter and Dymond, 1993) and have an important place in GIS. Radar imaging geometry varies with topography and high-resolution topographic data make a very significant contribution to the interpretation of radar image data (Kwok *et al.*, 1987). DEMs are also used for geometric correction of optical remotely sensed data (Bertolo *et al.*, 1993). Recently, software is becoming available to enable the use of DEMs to create an ortho image by differential rectification. Topographic features can have a pronounced effect on the radiometric response (Jones *et al.*, 1988) and these effects are often the cause of error in the automatic classification of digital imagery. DEMs can be used to correct for these effects. Conversely, the generation of DEMs directly from remotely sensed data (e.g., interferometric SAR or altimeters) appears very promising (Trotter and Dymond, 1993). For standard image geometric corrections, ground control points (e.g., road junctions) could be extracted in front of a vector database and their coordinates are used to register the image (Hartnall, 1994). Vector datasets from maps can be overlaid on the images to verify correction. Perhaps with advances in pattern recognition and line following techniques, lines on images could be registered to roads, rivers, and railways in vector datasets for image registration. Scene correlation could be performed using control areas and lines as well as points. Whether this will become operational reality is unknown, but an integrated raster and vector system would be an important requirement.

2.3.2 *Error circularity and amplifications*

One of the main dangers associated with the use of GIS data to assist in the analysis of remote sensing data is the propagation of error. Errors in remote sensing data appear in the form of misclassified or inappropriately labeled pixels or image segments arising from poor performance of the classifier and insufficient spectral or spatial resolution. With GIS data, the errors may be in position or attribute and may arise from lack of detail in the initial survey which produced the dataset together with subsequent inadequate interpolation. Errors such as inclusions of unresolved spatial units in bigger units and "slivers" between mapped polygons because of small errors in boundary mapping or digitization are common. When GIS datasets and remote sensing datasets are combined to make a new product, the different types of error will be propagated into the product. In some cases, when deterministic operations are carried out between spatial datasets, it is possible to construct effective models of the errors and their propagation (e.g., Heuvelink *et al.*, 1989; Heuvelink and Burrough, 1993; Fisher, 1991). However, in situations where the combination process involves both numerical and symbolic reasoning such as in a knowledge-based system, the possibility for effective modeling of error propagation is much reduced. The whole subject of error propagation in knowledge-based or expert systems is still poorly understood. This is undoubtedly an important area for further research on account of the benefits to be derived from the use of knowledge-based systems for integrated analysis of remote sensing and GIS datasets.

2.3.3 *Format conversion issues*

As reported in Section 2.1, many authors have presented results of environmental analyzes using GIS and remotely sensed data. However, virtually all have been forced to convert data to raster format at some stage in the analysis because the required fully integrated capabilities were not available in the systems used. This has raised a number of considerations. Gagliuso (1991) used GIS primarily for distance measures relative to habitat characteristics and human disturbance and rasterized a vector habitat map for use in the analysis. They state that the results of the distance analysis are affected by the choice of pixel size. This critical consideration, described also by Carver and Brundson (1994), is only necessary because the habitat map could not be used in its original vector format. Fung and Jim (1993) present examples of the use of remotely sensed and vector map data together in a raster image processing system for a study of hillfire locations. While the vectors could be displayed together with the image, they would have to be converted to raster for inclusion in further analysis. The conversion between formats is often slow and complex (Baumgartner and Apfl, 1994) and rasterizing creates error, i.e., measurable differences between input and output data. Most of the errors relate to the interpretation of which side of a linear boundary a pixel should be allocated when it physically straddles the boundary. Differences in results of conversion between systems, as described by Piwowar *et al.* (1990), occur and are not acceptable, especially when increasing amounts of data are exchanged between people and systems (van der Knaap, 1992). Baumgarter and Apfl (1994) believe that real advancements in the integration of remote sensing and GIS can only come with the development of "fully integrated" (Ehlers *et al.*, 1989) systems where such conversions are rarely necessary.

2.4 GIS and GPS

Geography Information System (GIS) is based on a geospatial database, with the support of computer software and hardware, to collect, input, manage, edit, query, analyze, simulate, and display spatially related data and adopt spatial model analysis methods to provide a variety of spatial and dynamic information in a timely manner for geographic research and decision making services and build up It is a computer technology system for geographic research and decision making. Global Navigation Satellite System (GNSS) is a space-based radio navigation and positioning system that can provide users with all-weather three-dimensional coordinate and velocity and time information anywhere on the Earth's surface or in near-Earth space. GIS can use GNSS technology to acquire spatial positioning data in real time, establish communication connections with GNSS receivers, and complete data processing and analysis in real time in order to realize the integration of GIS and GNSS and complete data acquisition, real-time positioning and navigation, real-time monitoring and other functions.

2.4.1 *GNSS service GIS data collection*

Nowadays, Geographic Information System (GIS) has been rapidly developed in various industries, such as urban planning, land management, communication facilities management, utilities, transportation, electric power, environmental protection, petroleum, forestry, and many other industries or systems. In these industries, GIS provides spatial data collection, storage, management, operation, and statistical summary and reporting to help decision-making. GIS data collection is the basis of GIS system applications; data collection efficiency has become the bottleneck of GIS analysis, maintenance, and update. GNSS is rapidly becoming a very important means of data acquisition for the GIS industry.

2.4.1.1 *GNSS RTK data acquisition*

With the rapid progress and popularization of GNSS technology, its role in field data measurement has become more and more important. Currently, Continuously Operating Reference Stations (CORS), a system of continuously operating reference stations established using multibase station network RTK technology (Figure 2.1), has become an important means of data acquisition. CORS is one or several fixed, continuously operating GNSS reference stations, using computers, data communications, and the Internet. CORS is a system that automatically provides different types of verified GNSS observations (carrier phase and pseudo-range), various correction numbers, status information, and other related GNSS service items to different types of users with different needs and levels in real time through a network consisting of a computer, data communication, and Internet (LAN/WAN) technologies. data transmission system, and user application system. CORS will provide dynamic real-time GNSS positioning services with centimeter-level accuracy for meteorology, vehicle and ship navigation and positioning, object tracking, public security and fire-fighting, mapping, GIS applications, which will greatly accelerate the construction of space infrastructure.

Real-Time Kinematic (RTK) is a measurement method that can obtain centimeter-level positioning accuracy in the field in real time (Figure 2.2), and its emergence has greatly improved the efficiency of field operations. RTK technology fixes the errors of GNSS signals (including atmospheric effects, satellite ephemeris errors, satellite clock aberrations, etc.) and makes the positioning results more accurate. The deviation of the positioning signal is known by mounting a mobile base station on a reference point at a known location. RTK provides an extremely important method of acquiring spatial data in real time, dynamically and precisely, and is an important data source for GIS, which greatly expands the application fields and applications of GIS, and the main application fields are the acquisition of GIS spatial base data; the local revision of terrain data traffic. The main application areas are as follows: the collection of GIS spatial

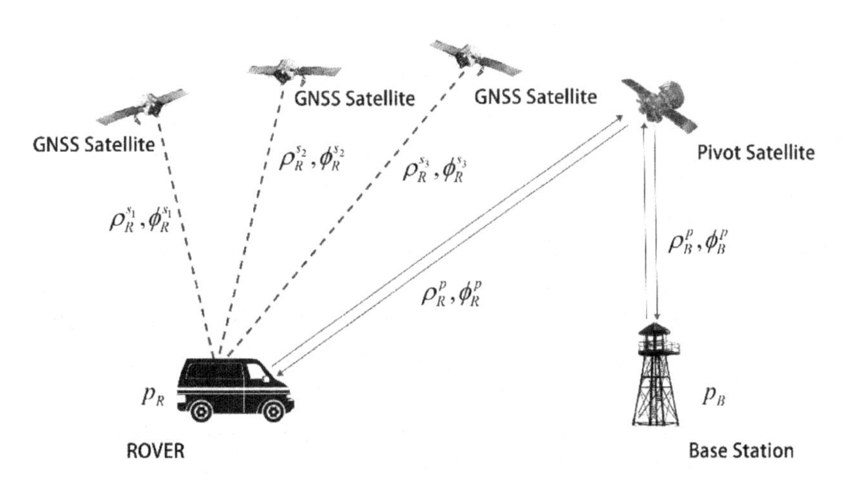

Figure 2.2. RTK-principle with code and phase observations.

basic data, the local revision of terrain data, the collection and updating of traffic data, the collection of periodic data, the provision of real-time positioning information for GIS, etc.

With the popularity of automated vehicles like drones and unmanned vehicles, the integration of RTK and vision localization technologies has become a trend. Vision enhanced RTK takes advantage of the mutual independence and exclusivity of GNSS and vision system (e.g., camera) data in terms of error sources (Figure 2.3) and deeply integrates high-precision GNSS positioning and computer vision technology, allowing the two to correct each other at the data layer, thus ensuring centimeter-level The accuracy of positioning is guaranteed in various complex scenes (woods or clectromagnetic interference environment).

Figure 2.3. Leica GS18 I.

2.4.1.2 *GNSS real-time positioning-assisted data acquisition*

(1) Aerial remote sensing data acquisition: Aerial remote sensing data acquisition mainly relies on aircraft or UAVs to carry out, by carrying airborne laser scanners, aerial digital cameras, directional positioning system POS (including GNSS and inertial navigator IMU), etc., to efficiently obtain point clouds, remote sensing images, and other data, among which aerial survey UAVs are widely used, changing the operation mode of data acquisition and greatly improving work efficiency. In areas where the terrain is more complex and information collection is more difficult, its flexibility is stronger. The flight control system is the core of the UAV to achieve autonomous flight control, which has an important impact on the stability, reliability of data transmission, position accuracy, and real-time of the UAV and plays a decisive role in its flight performance. In the flight control system, the GNSS

receiver as the most important sensor provides real-time position, maneuvering direction, travel speed, and time information for the UAV.

(2) Ground LiDAR data acquisition: LiDAR technology can provide a new technical means for the acquisition of spatial three-dimensional information data, using non-contact active measurement methods to quickly collect and obtain high-resolution data of target objects as well as spatial three-dimensional coordinate data. The ground LiDAR data acquisition is mainly divided into vehicle-mounted, ground-based, and handheld, and the instrument platform is equipped with laser scanner, panoramic image, high-precision GNSS equipment, etc., which can obtain high-precision positioning information in real-time, collect high-resolution panoramic image and high-precision 3D point cloud data, and provide data support for electric power patrol, forestry survey, mining measurement, underground space information acquisition, building façade measurement, BIM, and other fields. It provides data support for power patrol, forestry survey, mining measurement, underground space information acquisition, building facade measurement, BIM, and other fields.

2.4.2 *GNSS real-time location service GIS spatial analysis*

2.4.2.1 *GNSS service GIS big data system*

GNSS technology is becoming a public service and a new space–time infrastructure in the Internet of Things era. Nowadays, based on "Internet + Location (GNSS)", cloud computing, and big data technology, using "GNSS + Artificial Intelligence" fusion positioning, massive data access and storage, large-scale distributed computing and high concurrent real-time processing, and other core technologies, the air-sky integrated high-precision GIS big data system has been deeply applied in the fields of bicycle sharing, autonomous driving, smart phones, logistics monitoring, ship monitoring, etc., meeting the needs of the country, industry, and the public market for accurate location services.

2.4.2.2 *Smart driving applications*

High-precision positioning is an important part of intelligent perception in intelligent driving. GNSS INS high-precision positioning provides

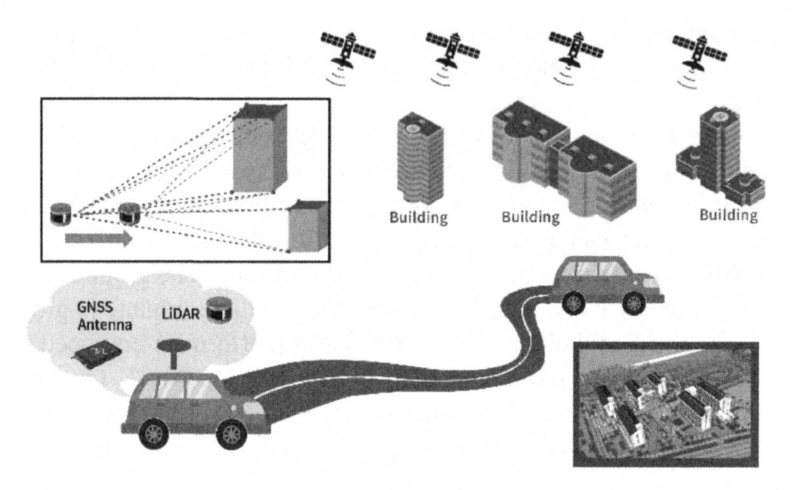

Figure 2.4. GNSS + GIS applied to autonomous driving.

real-time high-precision position, speed, and time information for intelligent driving vehicles and devices and provides reliable measurement results for global path planning of vehicles, time synchronization of various sensors, intelligent parking, three-dimensional intelligent traffic, and other needs by means of Knotai high-precision maps, high-speed communication, and cloud computing. GNSS positioning module or chip, as the core component of vehicle information system, provides position, speed, and time information, and GNSS positioning speed measurement performance determines the user experience of related functions, especially map navigation (Figure 2.4). With the increasingly complex urban road environment, such as dense road network, tall buildings, increased three-dimensional traffic, and more common glass curtain walls, higher requirements are put forward for GNSS technology and products, while the highly intelligent body makes the electromagnetic environment in the car more complex and also requires GNSS products to have better anti-interference capability and reliability.

2.4.2.3 *Smart city application*

Along with the continuous improvement of the mechanization of sanitation operations, the operation management of operational vehicles, business operation status, cost fine management and scientific assessment, and other higher requirements, in order to achieve efficient management transformation of sanitation vehicle business, the formation of a complete sanitation security vehicle online supervision system, the use of GNSS technology to obtain the dynamic location of sanitation operations vehicles, the use of GIS technology to build sanitation security vehicle monitoring platform, realizing manual management to information management, rough management to refined management. It provides a support platform based on modern information technology for building a fine management system of urban sanitation, optimizing the management concept of urban environmental sanitation, and improving the level of sanitation management and public services.

The priority of public transportation has become the development strategy of urban transportation, but the overall service level of the real public transportation system is still low, there are still problems such as insufficient overall carrying rate and low attractiveness, and there is still a gap between the services provided by the public transportation system and people's travel needs. Combining GNSS technology with 5G wireless communication technology, GIS technology, big data analysis, cloud computing, automatic control, and other advanced technologies, to create a GNSS-based integrated bus information dissemination system for the operating characteristics of bus vehicles, to improve the real-time bus forecast rate, greatly facilitates the public's bus travel.

2.4.2.4 *Wisdom agriculture application*

The promotion of mechanized operation in agricultural production has greatly improved the production efficiency and brought agricultural production from the operation method to the efficient mechanized operation. The application of GNSS enables agricultural machines to locate precisely in the field; the development of automatic control technology enables agricultural machines to operate precisely in the designated position; the development of GNSS technology and product promotion further promote the development of "precision agriculture". With the development of Internet and Internet of Things, and the wide application of various sensor technologies and products, agricultural development has entered the era of "smart agriculture", and GNSS has become the core technology for "smart agriculture" to provide accurate outdoor location.

2.4.2.5 *Smart logistics application*

In recent years, the distribution and express business in the logistics industry has become more and more important in the overall business. In the process of logistics production and operation, it is necessary to monitor and dispatch the personnel of the delivery vehicles to improve the production efficiency, on the one hand, and realize the tracking of goods and improve the service quality, on the other hand. Therefore, it is necessary to establish an informatization, graphical, and networked mobile operation service platform to meet the scheduling and monitoring of vehicles and personnel, improve efficiency, effectively reduce the waste of resources, and reduce costs; it can also enable the majority of customers to enjoy more rapid and timely services and meet the needs of mobile production operations such as mobile data exchange between personnel. The effective combination of GIS, GNSS, and wireless communication technology, supplemented by vehicle route model, shortest path model, network logistics model, distribution collection model, and facility positioning model, can establish a powerful logistics information system to make logistics real time and cost-optimized.

2.4.3 *GNSS high-precision positioning service for GIS monitoring system*

GNSS high-precision positioning service is used to monitor various facilities and features in real time and dynamically, and the information is connected to GIS monitoring system to realize dynamic monitoring of water conservancy and hydropower dams, high-rise buildings, large bridges, and hidden spots of geological disasters.

2.4.3.1 *Deformation monitoring*

Hydraulic hydropower dams as a large hydraulic building, its investment and after the completion of the effect is huge, but also due to its structure, operation environment, and other factors of complexity, coupled with the design, construction, operation, and maintenance of uncertainty, if the accidental deformation, the disaster caused by the wreck is also extremely serious. Therefore, continuous real-time monitoring of the operational status of water conservancy and hydropower dams is necessary, not only to provide safety assessment of the dam and ensure the safe operation of the dam but also to accumulate valuable technical information for the design and construction of similar projects. In 1998, China's diaphragm dam external deformation is for the first time using GNSS automated monitoring system (Figure 2.5). GNSS technology is a high-tech technology, which uses satellite technology for all-around measurement. The high accuracy of the technical parameters such as time information and 3D coordinates that it can provide is of great significance to the field of measurement. In order to ensure the safe operation of dams and reduce the occurrence of safety accidents, the degenerative factors of dams need to be analyzed and monitored in real time. Using GNSS to monitor the deformation of dams is the most applied technology nowadays, with the advantages of all-weather measurement, fast positioning, continuous real-time, and high automation.

With the development of national economy, more and more mega bridges have come into people's view, and the bridge structure has gradually developed into light and slender, while the load, span, and deck width of bridges have been growing and the structure type has been changing. These transportation infrastructures play a huge role in the economic life

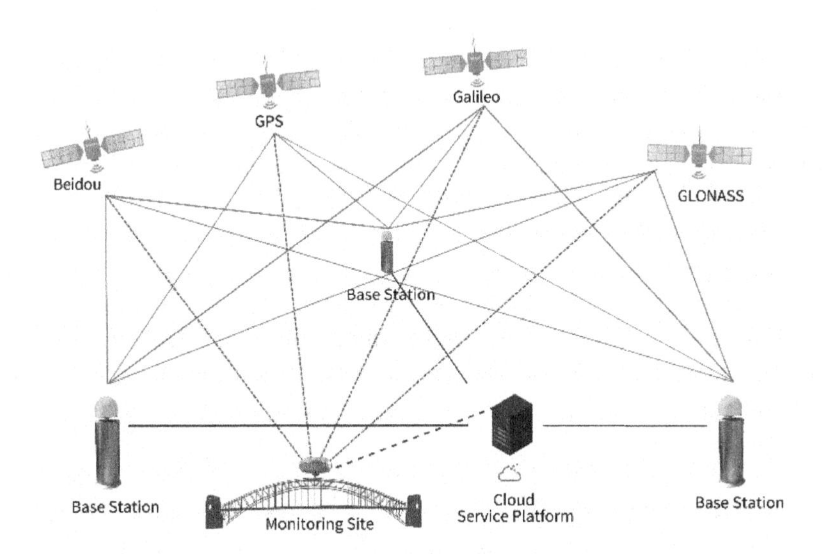

Figure 2.5. GNSS + GIS applied to deformation monitoring.

of people. Long-term bridge health monitoring is an important part of assessing the health status of bridges, ensuring bridge safety, and avoiding accidents. Monitoring bridge deformation can effectively reflect the working condition of bridge structures and provide a reliable data basis for the corresponding management departments. Therefore, providing early warning of bridge health condition with deformation monitoring is a work that must be carried out for a long time and is of great significance. GNSS technology can be used to obtain real-time changes of GNSS monitoring point displacements in characteristic parts such as bridge decks in real time, assess the health condition and structural safety of bridge structures through data connection with GIS monitoring system, construct 3D models to display the bridge displacement changes, and obtain comprehensive information on bridge operation conditions, which can be used to assess the safety, durability, and serviceability of structures.

2.4.3.2 *Geohazard monitoring*

With the continuous mining of large metal/non-metal open-pit mines, the drainage field, as a kind of giant artificial loose pile mat, is bound to have serious safety problems as its height keeps increasing. The destabilization of the drainage field will lead to mine soil disaster and major engineering accidents, which will not only affect the normal production of the mine and personal safety of employees but also directly lead to huge economic losses for the mine. Combining with the current situation of the mine and hydrogeological conditions, GNSS+sensor monitoring points are reasonably laid out to form a complete monitoring profile and monitoring network, which becomes a systematic and three-dimensional deformation monitoring system and can timely grasp the deformation of the slope accumulation area of the soil drainage field, analyze the deformation law, predict the range of possible changes of the slope and landslide and its change trend, make timely and rapid evaluation of the current situation of the landslide area and the affected area, make prediction and forecast, and provide decision basis for management personnel and related experts.

For many years, China has been suffering from the impact of geological disasters, and both the safety of people's lives and the safety of national property are at great risk, which mainly includes forms including landslides, mudflows, collapses, ground subsidence, and cracks. Therefore, it is urgent to strengthen the monitoring and prevention of geological disasters. At present, the means of geological disaster monitoring in China mainly adopts group measurement and group prevention, and professional monitoring is also mainly manual monitoring, with fewer automated monitoring points, and even fewer monitoring points using remote transmission control, so the overall monitoring efficiency is not high. Landslide monitoring includes overall deformation monitoring of the landslide body, stress–strain monitoring of the landslide body, external environment monitoring such as rainfall, and groundwater level monitoring. Landslide monitoring requires a variety of techniques and methods, of which deformation monitoring is an important element and an important basis for determining landslides. In the past, the method of deformation monitoring was to use conventional geodetic methods, i.e., total station conductor and electromagnetic wave ranging triangle elevation methods. The above methods have a large operational workload and are difficult to operate in mountainous areas.

After the emergence of GNSS technology, because GNSS positioning is to use the reception of airborne satellite signal ranging for positioning, experts and scholars at home and abroad have shown that applying the precision ephemeris of International GNSS Geodynamics Service (IGS) and GAMIT (GNSS data processing software) to process data, the short and medium sides are better than 1.4×10^{-7} relative to the medium error and the long side is better than GNSS and has the advantages of high accuracy, not subject to weather interference, continuous and long time real-time monitoring, good automation, and easy integration, which is especially suitable for geological disaster monitoring. It is especially suitable for geological hazard monitoring. Since GNSS monitoring stations do not need to communicate with each other, it greatly reduces the workload. Moreover, GNSS observation data can be uploaded to the indoor data processing center using wireless communication technology and integrated with GIS system to realize dynamic monitoring of geological hazards.

2.5 Future Prospects for Remote Sensing Products and GIS Implications

2.5.1 *Developments in sensors and modes of use*

Even though, as we have seen, many problems still remain in the integration of GIS and remote sensing data, the characteristics of the datasets in question are constantly evolving and presenting new opportunities and new challenges. It can be argued that remote sensing entered a new generation with the launch of the operational SAR system on the ERS-1 satellite in 1991. The ERS-1 SAR is capable of providing high-resolution imagery of the Earth's surface in all weathers and its images can be easily integrated with data from the Landsat-TM. This provides new datasets describing not only the reflectance of visible and near-infrared radiation and the emission of thermal infrared radiation from the surface but also the backscattering of the surface in the microwave region of the electromagnetic spectrum. A new dimension has therefore been added to remote sensing which offers a much enhanced capability for environmental monitoring and mapping and, as a consequence, for generating new GIS datasets. For example, experiments on mapping land cover have demonstrated that accuracy can be improved significantly by use of SAR data in addition to TM (Wilkinson *et al.*, 1993) since the SAR backscatter signal can be much more sensitive than TM to the roughness and structure of surface features, rocks, vegetation, etc. The continuity of SAR imagery is assured in the foreseeable future with satellites such as JERS-1, Radarsat, and ERS-2. The integrated use of SAR data and conventional visible/infrared data is still in its infancy, however. At the end of the current decade, however, remote sensing will enter yet another generation with the launch both of ENVISAT and of the polar platforms which will comprise the Earth Observing System. Several important technological advances will be made for land mapping including the use of the following:

(a) Medium-resolution imaging spectrometers (i.e., the MODIS and MERIS systems on the NASA and ESA polar platforms). These will have 36 separate spectral channels, 250–1000 meter ground resolution, and radiometric calibration of unprecedented accuracy (Barker, 1993).
(b) Multiple look-angle sensors (e.g., the Multi-Angle Imaging Spectroradiometer, "MISR", Diner *et al.*, 1991).

In addition, the future SPOT-5 satellite may carry a sensor with a ground resolution of the order of 2.5–5 m (Novajosky, 1993). This will also be launched towards the end of the decade constituting a further landmark in the evolution of remote sensing. CEOS (1992) provides a comprehensive survey of the remote sensing systems planned for the next 10 years.

2.5.2 *GIS implications of changes in sensors and research needed*

The rapid developments now being made in remote sensing systems have some important implications for the integration of remote sensing and GIS data. As the datasets provided by remote sensing become more complex, the derived products will become less and less like maps and flat two-dimensional plots representing the landscape will become less useful. The new remote sensing datasets, especially with many spectral channels and with frequent repeat coverage, will contain large numbers of spatially coincident data layers acquired on different dates. The possibilities for undertaking diagnostic studies on the state of the environment are inestimable at the present time. It is clear, however, that the utilization and exploration of such datasets in combination with pre-existing GIS data cannot be achieved efficiently with the current state of GIS technology (see Table 2.1). As long as the integrated use of a remote sensing dataset and a GIS dataset only involves a small number

Table 2.1. Emerging trends in the gathering and integrated analysis of remote sensing and GIS data.

Attribute Analysis/Classification of Remote Sensing and GIS Data
• Artificial neural networks
• Knowledge-based systems
• Integrated statistical and neural systems
• Improvements in uncertainty handling (inc. “fuzzy” method)
Use of Very High Dimensionality Data from Remote Sensing and GIS
• Advanced visualization (inc. “Virtual Reality”)
• Dimensionality reduction tools (e.g., “projection pursuit”)
Remote Environmental Mapping
• Hyperspectral techniques (image spectrometry)
• Fusion of multisensor, multiview angle, multidate data (optical, IR, SAR)
Object analysis/Search in Integrated Remote Sensing and GIS Datasets
• Self-organizing neural networks
• Integrated spatial and temporal representation and analysis tools
• Exploration tools (data “mining”)

Source: Wilkinson (1996).

of data layers which are processed by techniques such as sieve mapping, the existing GIS technology based on two-dimensional digital maps is usable — even if error propagation is still a serious and largely unsolved problem. However, with the currently foreseen data explosion, new approaches are needed. It is possible to highlight a few of the approaches which are likely to offer much promise in the future (Figure 2.6) such as "virtual reality" (VR) technology and multimedia systems.

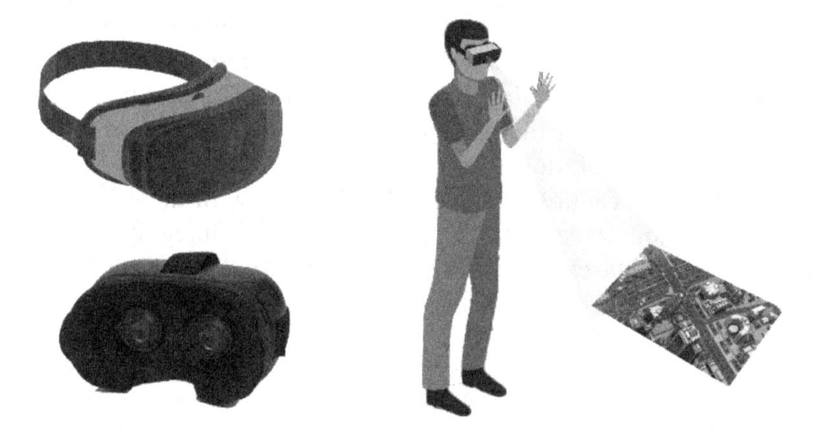

Figure 2.6. Virtual reality glasses provides a touch-less interaction to manipulate a given 3D scene.

Apart from its use for computer entertainment, virtual reality is now growing in its application to computer graphic applications where there is a need for the user to enter and explore a three-dimensional visual world (Ellis, 1993). In order to explore integrated remote sensing and GIS datasets of very high dimensionality, the use of VR is likely to become indispensable alongside developments in 3D visualization techniques (e.g., Cetin *et al.*, 1993; Eddy and Looney, 1993; Keim and Kriegel, 1994). However, although 3-D viewing will make a significant improvement compared to flat 2-D displays, the problem of how to reduce very high dimensional datasets to three dimensions will remain. To deal with this problem, it will be necessary to develop methods such as "Projection pursuit" (Jones and Sibson, 1987) which is beginning to be used in remote sensing (Jimenez and Landgrebe, 1994).

Apart from new visualization tools based on VR, there will be a need for much improved spatial analysis tool to find and extract patterns from the new datasets. It is likely that neural networks will play an important role in this task and that there will need to be better implementations of such algorithms in parallel computer hardware to deal with very large datasets. Certainly, unsupervised data clustering procedures such as self-organizing neural networks (e.g., Kohonen, 1984) will have important uses. Also, new techniques for exploratory data analysis (e.g., Anselin, 1994) and methods for data mining for searching very large spatial databases for meaningful patterns are potentially interesting but need much more development. Besides improving our ability to search and analyze complex spatial data, it should not be overlooked that most datasets captured from space have a temporal aspect and that most features of the natural environment are constantly evolving. Improved models are undoubtedly needed for handling time in GIS (Worboys, 1994) and for linking temporal changes or signatures revealed in sequences of remote sensing data to the evolution of objects on the landscape. It is overwhelmingly apparent that the trends in data capture and data collection are rapidly exceeding the pace of developments in GIS technology and data analysis techniques both in the spatial and temporal domains. This trend needs to be appreciated by GIS developers.

2.5.3 *Terminology*

The most important consideration in selecting a system for environmental analysis is whether it will allow the user to make the most of the data available and help improve the ability to make decisions and conclusions. If it does, then clearly labels do not matter. But should terminology be formalized? Baumgartner and Apfl (1994) described "an integrated geographic information system with remote sensing, GIS, and consecutive modeling" and then state that they "do not aim at a totally integrated system as described by Ehlers *et al.* (1989)". While integrated GIS has been the subject of research initiatives (Star *et al.*, 1991), comments (Dobson, 1993), and research (Jackson and Mason, 1986) for some years, users who are actually attempting to do this have had to use inappropriate software, convert formats, and transfer datasets between systems (Trotter, 1991). Nellis *et al.* (1990) use the term "interface" (n. a means or place of interaction between two systems) which is, for many of the examples cited in this paper, a more accurate description of the relationship between remote sensing and GIS they employ. However, the most important concern is to get GIS users use remotely sensed data and remote sensing scientists using GIS in order to extract the maximum information from the data. It could be argued that the best way to achieve this is to allow users to use the best format for the data and to provide functionality to enable them to integrate their datasets with ease. This would be substantially improved by the "fully integrated" approach of Ehlers *et al.* (1989). Their definition, however, could be refined with a proviso. An integrated GIS must not rely on data conversion for effective integration and data flow between the images and the vector database must be two-way.

Perhaps, the phrase "Integrated GIS" is (or should become) a tautology. Remote sensing data are after all a form of geographical information. We should hope that all users of geographical information will expect to include EO data collected from aircraft and as a matter of course expect to be able to use it seamlessly with their other datasets. This is an aim of the British National Space Centre which has sponsored a number of projects in this field (e.g., Brown and Fletcher, 1994; Hinton, 1995a). When users also take this line, fully integrated system development will be user led.

2.5.4 *Future directions*

The trend in remote sensing over the past decade has been from empirically based image classification, mapping, and inventory to more deterministic modeling of scene characteristics based on physical laws of radiative transfer and energy balance and to knowledge-based image interpretation systems (Davies *et al.*, 1991). GIS analyzes have moved from simple map overlay and relational models to spatially distributed simulation modeling. Future developments in remote sensing depend critically on hardware and software that help the integration of remote sensing and GIS. Improvements in interfaces between image processing, GIS, database management, expert systems, and statistical software will go a long way in improving analysis capabilities (Davis *et al.*, 1991). However, in describing a potential "fully integrated" image processing and geographical information system, Ehlers *et al.* (1989) said that "the final system would have more capabilities than just the sum of the two". The ability to use vector objects to constrain the query and examination of raster data is a valuable development in GIS and one which a fully integrated system design can exploit.

Future requirements for remote sensing and GIS integration were discussed by Davies *et al.* (1991) and some priorities were identified. Systems must include modeling tools which are discipline specific at least in terminology and parameters used and have the ability to cope with data of different levels of quality and to report on validity of results produced. Software should integrate seamlessly with existing report generating software such as spreadsheets, word processors, databases, and graphics packages and must take advantage of the advances in network computing. Developments of integrated GIS must also consider the new image sources becoming available, particularly the very high spatial (1–3 m) and spectral resolution imagery, including interferometric and high-resolution radar. These new data sources require processing functionality not currently standard in many systems (Poulter, 1995).

The next 10 years promise to be very exciting in remote sensing, yet, at the same time, there is a current sense that many fundamental problems still remain to be solved in making really effective use of spatial data, however it is acquired. These problems are fundamental that there is, for example, still little consensus over what kinds of data structures to use or how to manage error and uncertainty. Conventional cartographic datasets encoded in digital form in GIS will always form the bedrock of our analysis of the spatial world. But on top of this, we will soon have datasets of enormous dimensionality from remote sensing revealing patterns and complexity in the natural environment that is difficult to imagine at the present time. The need for new tools and techniques to handle these integrated datasets is considerable.

The next step is perhaps integration with expert systems. Trotter (1991) states that image understanding can involve the use of prior knowledge often held at different levels of generalization to the image data. Expert systems are considered to have significant potential for providing a general approach to the routine use of ancillary data in image classification (Goodenough *et al.*, 1989; Hartnett *et al.*, 1994) although they remain experimental. In some parts of the world where digital cartographic information is scarce, techniques for extracting such structural information from EO images based on expert knowledge would be a valuable advance (Hinton, 1995b). The future for integrated Geographical Information Systems is challenging but their continued development will enable EO data to be used operationally in a wide range of environmental applications.

References

Abel, D. J., Kilby, P. J., and Davis, J. R. (1994). The systems integration problem. *International Journal of Geographical Information Systems*, **8**(1): 1–12.

Archibald, P. D. (1987). GIS and remote sensing data integration. *Geocarto International*, **2**(3): 67–73.

Arduini, F., Berta, R., Fioravanti, S., and Giusto, D. D. (1991). Texture characterization and SAR image segmentation. In *Proceedings of the 7th Workshop on Multidimensional Signal Processing*, September, pp. 9–16. IEEE.

Bajcsy, R. and Tavakoli, M. (1976). Computer recognition of roads from satellite pictures. *IEEE Transactions on Systems, Man, and Cybernetics*, (9): 623–637.

Baumgartner, M. F. and Apfl, G. (1994). Towards an integrated geographic analysis system with remote sensing, GIS and consecutive modelling for snow cover monitoring. *International Journal of Remote Sensing*, **15**(7): 1507–1517.

Burrough, P. A. (1986). *Principles of Geographical. Information Systems for Land Resource Assessment*. Oxford: Clarendon Press.

Chagarlamudi, P. and Plunkett, G. W. (1993). Mapping applications for low-cost remote sensing and geographic information systems. *International Journal of Remote Sensing*, **14**(17); 3181–3190.

Chiang, K.-W., Tsai, G.-J., Li, Y.-H., Li, Y., and El-Sheimy, N. (2020). Navigation engine design for automated driving using INS/GNSS/3D LiDAR-SLAM and integrity assessment. *Remote Sensing,* **12**: 1564.

Cleynenbreugel, J. V., Fierens, F., Suetens, P., and Oosterlinck, A. (1990). Delineating road structures on satellite imagery by a GIS-guided technique. *PE&RS, Photogrammetric Engineering & Remote Sensing*, **56**(6): 893–898.

Chiou, W. C. (1985). NASA image-based geological expert system development project for hyperspectral image analysis. *Applied Optics*, **24**(14): 2085–2091.

Dayhoff, J. E. (1990). Regularity properties in pulse transmission networks. In *1990 IJCNN International Joint Conference on Neural Networks*, June, pp. 621–626. IEEE.

De Cola, L. (1989). Fractal analysis of a classified Landsat scene. *Photogrammetric Engineering and Remote Sensing*, **55**(5): 601–610.

Densham, P. J. (1994). Integrating GIS and spatial modelling: visual interactive modelling and location selection. *Geographical Systems*, **1**(3): 203–219.

Dobson, J. E. (1993). The geographic revolution: A retrospective on the age of automated geography. *The Professional Geographer*, **45**(4): 431–439.

Ehlers, C. L., Wall, T. L., Wyss, S. P., and Chaplin, R. I. (1989). Social zeitgebers: A peer separation model of depression in rats. In *Animal Models of Depression*, pp. 99–110. Boston: Birkhäuser.

Ehlers, M., Edwards, G., and Bedard, Y. (1989). Integration of remote-sensing with geographic information-systems — A necessary evolution. *Photogrammetric Engineering and Remote Sensing*, **55**(11): 1619–1627.

Faust, N., Anderson, W. and Star, J. (1991). Geographic information systems and remote sensing future computing environment. *Photogrammetric Engineering and Remote Sensing*, **57**(6): 655–668.

Franklin, S. E. and Peddle, D. R. (1990). Classification of SPOT HRV imagery and texture features. *Remote Sensing*, **11**(3): 551–556.

Gagliuso, R. A. (1991). Remote sensing and GIS technologies: An example of integration in the analysis of cougar habitat utilization in southwest Oregon. *GIs Applications in Natural Resources*, pp. 323–329.

Goodenough, D. G., Goldberg, M., Plunkett, G., and Zelek, J. (1987). An expert system for remote sensing. *IEEE Transactions on Geoscience and Remote Sensing*, (3): 349–359.

Goodchild, M., Haining, R., and Wise, S. (1992). Integrating GIS and spatial data analysis: problems and possibilities. *International Journal of Geographical Information Systems*, **6**(5): 407–423.

Green, K. (1992). Spatial imagery and GIS. *Journal of Forestry*, **90**(11): 32–36.

Haralick, R. M., Shanmugam, K., and Dinstein, I. H. (1973). Textural features for image classification. *IEEE Transactions on Systems, Man, and Cybernetics*, (6): 610–621.

Heßelbarth, A., Medina, D., Ziebold, R., Sandler, M., Hoppe, M., and Uhlemann, M. (2020). Enabling assistance functions for the safe navigation of inland waterways. *IEEE Intelligent Transportation Systems Magazine,* **12**: 123–135.

Hutchinson, C. F. (1982). Classification improvement. *Photogrammetric Engineering and Remote Sensing*, **44**(1): 123–130.

Hinton, J. C. (1996). GIS and remote sensing integration for environmental applications. *International Journal of Geographical Information Systems*, **10**(7): 877–890.

Johansen, M. E., Tommervik, H., Guneriussen, T., and Pedersen, J. P. (1994). Using a GIS (ArcInfo) as a tool for integration of remote sensed and in-situ data in an analysis of the air pollution effects on terrestrial ecosystems in Varanger (Norway) and Nikel-Pechenga (Russia). In *Proceedings of International Geoscience and Remote Sensing Symposium*, Vol. 2, pp. 1213–1216.

Kimes, D. S., Harrison, P. R., and Ratcliffe, P. A. (1991). A knowledge-based expert system for inferring vegetation characteristics. *Remote Sensing*, **12**(10): 1987–2020.

Kimes, D. S. and Holben, B. N. (1992). Extracting spectral albedo from NOAA-9 AVHRR multiple view data using an atmospheric correction procedure and an expert system. *International Journal of Remote Sensing*, **13**(2): 275–289.

Kontoes, C., Wilkinson, G. G., Burrill, A., Goffredo, S., and Megier, J. (1993). An experimental system for the integration of GIS data in knowledge-based image analysis for remote sensing of agriculture. *International Journal of Geographical Information Systems*, **7**(3): 247–262.

Kwok, R., Curlander, J. C., and Pang, S. S. (1987). Rectification of terrain induced distortions in radar imagery. *Photogrammetric Engineering and Remote Sensing*, 53(5): 507–513

Lv, Z. and Li, X. (2016). Virtual reality assistant technology for learning primary geography. In Zhiguo Gong, Dickson K. W. Chiu, and Di Zou (Eds.), *Current Developments in Web Based Learning*. [Online]. Cham: Springer International Publishing. pp. 31–40.

Matsuyama, T. (1987). Knowledge-based aerial image understanding systems and expert systems for image processing. *IEEE Transactions on Geoscience and Remote Sensing*, (3): 305–316.

Mattikalli, N. M. (1994). An integrated geographical information system's approach to land cover change assessment. In *Proceedings of IGARSS' 94-1994 IEEE International Geoscience and Remote Sensing Symposium*, August, Vol. 2, pp. 1204–1206. IEEE.

Merchant, J. W. (1994). GIS-based groundwater pollution hazard assessment: A critical review of the DRASTIC model. *Photogrammetric Engineering and Remote Sensing*, **60**: 1117–1117.

McKeown, D. M. (1987). The role of artificial intelligence in the integration of remotely sensed data with geographic information systems. *IEEE Transactions on Geoscience and Remote Sensing*, (3): 330–348.

Nicolin, B. and Gabler, R. (1987). A knowledge-based system for the analysis of aerial images. *IEEE Transactions on Geoscience and Remote Sensing*, (3): 317–329.

Pao, Y. (1989). Adaptive pattern recognition and neural networks.

Pentland, A. P. (1984). Fractal-based description of natural scenes. *IEEE Transactions on Pattern Analysis and Machine Intelligence*, (6): 661–674.

Pentland, A. P. (1985). On describing complex surface shapes. *Image and Vision Computing*, **3**(4): 153–162.

Perkins, D. N. (1986). Thinking frames. *Educational Leadership*, **43**(8): 4–10.

Piwowar, J. M., LeDrew, E. F., and Dudycha, D. J. (1990). Integration of spatial data in vector and raster formats in a geographic information system environment. *International Journal of Geographical Information System*, **4**(4): 429–444.

Priebe, C. E., Solka, J. L., and Rogers, G. W. (1993). Discriminant analysis in aerial images using fractal-based features. In *Adaptive and Learning Systems II*, September, Vol. 1962, pp. 196–208. SPIE.

Rumelhart, D. E., Hinton, G. E., and Williams, R. J. (1986). Learning representations by back-propagating errors. *Nature*, **323**(6088): 533–536.

Trotter, C. M. (1991). Remotely-sensed data as an information source for geographical information systems in natural resource management a review. *International Journal of Geographical Information System*, **5**(2): 225–239.

Wang, F. and Newkirk, R. (1988). A knowledge-based system for highway network extraction. *IEEE Transactions on Geoscience and Remote Sensing*, **26**(5): 525–531.

Wilkinson, F., Worrall, D. R., and Williams, S. L. (1995). Primary photochemical processes of anthracene adsorbed on silica gel. *The Journal of Physical Chemistry*, **99**(17): 6689–6696.

Worboys, M. F., Hearnshaw, H. M., and Maguire, D. J. (1990). Object-oriented data modelling for spatial databases. *International Journal of Geographical Information System*, **4**(4): 369–383.

Wu, J. K., Cheng, D. S., Wang, W. T., and Cai, D. L. (1988). Model-based remotely-sensed imagery interpretation. *International Journal of Remote Sensing*, **9**(8): 1347–1356.

Yu, J., Meng, X., Yan, B., Xu, B., Fan, Q., and Xie, Y. (2020). Global navigation satellite system-based positioning technology for structural health monitoring: A review. *Structural Control and Health Monitoring,* **27**: e2467.

Chapter 3

Parallel Geoprocessing

3.1 Introduction

In a more modest, but much more widespread fashion, the technology of symmetric multiprocessing, be it with Pentium, SPARC, or Alpha processors, has wrought a quiet revolution in the fileservers of innumerable academic institutions and commercial organizations. While the number of processors, typically 4, 8, or 16, is much smaller than with massively parallel processing, considerable gains in throughput can be achieved for medium-sized workgroups using this approach.

From the humblest Webserver to the most ambitious mirror site, Internet providers are also looking to parallel processing servers to address the exponential growth in traffic across the Information Superhighway.

However, in between the "Grand Challenge" problems of global science, the meanderings of a million Web browsers, and the contented office staff now connected to a multiprocessing fileserver, there remains a large group of analysts in the GIS and remote sensing field, struggling with large data volumes, processing bottlenecks, and limited throughput. Little or none of their standard software tools will run on massively parallel machines or have many of these tools yet been modified to utilize all the available power on symmetric multiprocessors.

A number of reasons can be given for this. First, the community, both vendors and users, is, of necessity, risk averse because of the inherent complexity of the algorithms and applications. Second, the lack of stability and standardization in parallel software until recently has not favored a field where most of the software technology is in the commercial rather than the scientific domain. This position is now changing markedly, with the arrival of Massive Parallel Language (MPL), the message passing standard for parallel software that insulates application programs from the underlying hardware. Third, very little systematic research has yet been reported in the literature into the impact of a parallel approach on existing GIS algorithms originally developed for serial machines of very limited processing capacity. Fourth, until large volumes of digital cartographic data became widely available, much of the GIS user's time was consumed in data automation rather than consideration of appropriate kinds of analysis. This situation is now changing rapidly, with user expectations rising, in terms of scenario analysis, links from the GIS to environmental or statistical modeling software, and integration of GIS with multitemporal remote sensing, to name but a few examples.

Given the long gestation time of new GIS software technology and the high cost of entry, it is difficult for vendors, never mind small research groups, to respond adequately to the rapid shift in user requirements towards sophisticated analysis of very large datasets. Realistically, only parallel processing technology can provide the kind of quantum leap in performance need to open up new avenues of investigation in such fields as dynamic visualization of spatial data, interactive multilayer overlay on large datasets, and real-time GIS. It is therefore of particular concern that so little fundamental research has been undertaken on parallel algorithms for GIS, at a time when such great opportunities are presenting themselves in the analytical domain.

The development of high-performance computing technology has had a significant impact on the process of science. In the GIS community, the importance of parallel processing for geoprocessing is increasingly recognized. Indeed, Dangermond and Morehouse (1987) suggest that parallel processing is one of the most significant trends in hardware for GIS. Although high performance often is cited as the primary benefit of and motivation for adopting parallel processing, the primary benefit that accrues to the geographer is improved understanding of how to represent space and spatial relations and how to exploit them in analytical procedures. Parallel processing forces us to address these issues in ways that serial computers do not. Thus, parallel geoprocessing offers two benefits: the opportunity to investigate new ways of representing, manipulating, analyzing, and exploiting space and spatial relations; and greatly increased computational throughput.

A number of researchers have investigated the potential of parallel and other forms of high-performance computing to enhance geoprocessing. Peucker and Douglas (1975), for example, designed three algorithms to extract terrain features in which the extraction processes, implemented as operations on neighborhood data, are independent of each other and can be run on a parallel processing computer. In the domain of transportation and land use modeling, Harris (1985) described possible methods for solving a doubly constrained gravity model and for finding least-cost paths using a massively parallel (SIMD — Single Instruction, Single Data) computer. Langran (1991) proposed using parallel software models to integrate the components of cartographic generalization. She argued that spatial problems, especially cartographic generalization, are suitable for parallel processing because of their multiple dimensions and very large data volumes. Taking a somewhat different tack, Dangermond and Morehouse (1987) discussed methods for solving problems in a pipeline fashion and how to take advantage of inherent spatial sub-divisions when partitioning spatial problems. In a similar vein, Healey and Desa (1989) compared parallel architectures and described parallel programming languages, three approaches to parallelizing algorithms, and the potentialities and problems of Transputer-based parallel processing for supporting GIS.

Other researchers have used vector supercomputers to improve the computational throughput of geoprocessing. Humphrey and Zoraster (1985), for example, described design issues for implementing cartographic algorithms on vector supercomputers. In contrast, Sandu and Marble (1987) evaluated the performance of spatial data handling and visualization algorithms that exploit the vectorization capabilities of a Cray X-MP supercomputer.

Research on the use of explicitly parallel computer architectures has focused on SIMD and Multiple Instruction streams, Multiple Data streams (MIMD) machines. Li (1992), for example, investigated the use of SIMD for map data analysis. He discussed methods for mapping different types of grid modeling into SIMD parallel systems and presented some initial results for a CM-2 supercomputer. Also working with CM-2 supercomputers, Mower (1992) developed a new parallel procedure to identify drainage basins from raster DEM and Puppo *et al.* (1994) designed and implemented a parallel algorithm to build a triangulated irregular network (TIN). Taking a broader view, Bestul (1991) presented a general approach for creating parallel algorithms on point-based quadtree data structures for a massively parallel (SIMD) architecture while Franklin *et al.* (1989)

implemented an algorithm for line intersection on a 16 processor SIMD computer that yielded speedup ratios of 8–13, depending on the data structure that was used.

Research with MIMD systems has tended to focus on GIS operations and methods of network analysis. Dowers *et al.* (1991) analyzed the performance of GIS on workstation-based, parallel architectures and concluded that, although the utilization of parallel processing will yield orders of magnitude increases in performance, it will require considerable investment in scarce and highly skilled programming resources. Verts and Thomson (1988) examined parallel approaches using associative memory: they showed that the performance of raster-based GIS query and database manipulation can be improved significantly using hardware parallelism. Working in a vector environment, Hopkins *et al.* (1992) presented results on the scalability of their polygon overlay algorithm in a distributed-memory MIMD system.

A recursive parallel algorithm for triangulation of unordered digital terrain models (DTMs) was introduced by Collins (1989). The algorithm decomposes the vector-based DTM into mutually exclusive blocks, triangulates each concurrently, and merges the sub-results by "knitting" them together. A number of researchers have used transputer arrays (distributed-memory MIMD systems); Hopkins and Healey (1990) developed a parallel algorithm for detecting line intersections; Ware and Kidner (1991) implemented Delaunay Triangulation; and Rokos and Armstrong (1992) implemented a parallel algorithm for terrain feature extraction.

Smith *et al.* (1989) designed an asynchronous parallel algorithm for the weighted region, least cost path problem. They partitioned the region into triangular sub-regions with equal costs of traversal, located a processor at each edge, and used local communication to and least cost paths. Working with distributed-memory MIMD systems, Ding *et al.* (1992) developed decomposition methods for shortest path algorithms and Armstrong and Densham (1992) described strategies for parallelizing location-allocation models.

To summarize, while early work on parallel geoprocessing focused on SIMD architectures, interest in MIMD computers has increased greatly in the last few years. Many issues concerning the implementation of GIS algorithms on MIMD architectures have been identified (Healey and Desa, 1989; Armstrong and Densham, 1992; Hopkins *et al.*, 1992) and strategies developed for the design and mapping of these algorithms to MIMD systems (Ding, 1993). It is also important, however, to study parallel concepts and paradigms that can be sued to design new and more efficient algorithms for geoprocessing.

Other areas of research also hold promise for parallel geoprocessing. Parallel matrix methods are applicable to spatial analysis and modeling (Quinn, 1987; Modi, 1988; Bertsekas and Tsitsiklis, 1989). Similarly, graph theory is widely applied to spatial problems in vector-based GIS and network analysis and the development of parallel graph algorithms attracts much research effort (Quinn and Deo, 1984). Image processing is a popular research topic, and themes include the following: basic algorithms for parallel and real-time image processing, image algebra, and MIMD-based applications (Preston and Uhr, 1982). Computer vision also requires intensive computation and has similarities with the visualization of spatial data in GIS (Chaudhary and Aggarwal, 1990). Furthermore, recent developments in parallel database systems suggest they have a very promising future. Although queries in relational databases are ideally suited to parallel processing, because they consist of uniform operations applied to uniform streams of data, the choice of architecture is very important (Dewitt and Gray, 1992).

3.2 Fundamental Issues in Parallel Geoprocessing

To realize the full potential of parallel geoprocessing, researchers must address several issues: how to measure the performance of parallel processing, how to identify the factors affecting performance, and how to achieve high performance.

3.2.1 *Parallel processing and parallel computer architectures*

Parallel processing is a class of computerized information processing that emphasizes the concurrent manipulation of data segments or concurrent execution of process components to solve a problem or to accomplish a task (Quinn, 1987), often on a specially designed computer. Using Flym's (1966) classification, common parallel computer architectures fall into two categories: Single Instruction stream, Single Data stream (SIMD) and Multiple Instruction streams, Multiple Data streams (MIMD). In SIMD systems, each of the processors executes simultaneously the same instruction set on its own dataset. In contrast, the processor in MIMD systems works simultaneously on different datasets using different instructions or tasks. If all the processors in a MIMD system execute the same instructions, however, they are simulating a SIMD system. Thus, MIMD architectures are more general purpose and flexible than SIMD architectures (Ding, 1993).

Parallel computers can be classified into shared-memory and distributed-memory systems. In shard-memory systems, all processors have equal access to all memory, called global memory. This eases programming because only the algorithm is parallelized and data are left undivided in global memory (Fox *et al.*, 1988). Memory often is accessed through a common system bus or a switching network; but as the number of processors increases, it can become a bottleneck that limits the speed at which processors can access memory (Ragsdal, 1991). The fundamental limitation of shared-memory systems is that individual memory modules can be accessed by only one processor at a time, so all but one of the processors seeking access to a given module will be blocked for the duration of an individual access.

In distributed-memory computers, each processor has its own private memory, called local memory. When an operation on one processor requires data located in another memory, an explicit message must be sent from one processor to another through an interconnecting network. This approach is more efficient if local memory access is more frequent than other types of access because message exchange time is proportional to message length (Denning and Tichy, 1990). By providing uniform and fast memory access times, distributed-memory systems offer higher absolute performance than shared-memory systems for many scientific applications (Ragsdale, 1991). Indeed, Denning and Tichy (1990) argue that designers will continue to favor distributed memory architectures because most parallel algorithms can be designed to localize the data dependency of each processor.

A processor's unit of computational work is called a grain (Denning and Tichy, 1990). A fine grain usually contains only one or a very few data elements while a coarse grain often has many. Grain size both depends on the characteristics of a parallel architecture and significantly affects its performance. Although all algorithms require communication time for input and output of data, only parallel algorithms require communication time to synchronize concurrent processing among multiple processors. Large grain sizes are favored because they reduce communication as a proportion of total processing time (Fox *et al.*, 1988).

3.2.2 *Performance analysis of parallel algorithms*

Two important performance measures for parallel algorithms are speedup and efficiency. Speedup is the ratio of the time required to complete a given computation on a single processor of the parallel computer to the time required for the same computation performed on N processors of the computer. If $T(N)$ denotes the time elapsed on a parallel computer with N processors, the sequential time, $T_{sequential}$, can be described as

$$T_{sequential} = T(1) \tag{3.1}$$

Thus, the speedup, S, with N processors is given by

$$S(N) = \frac{T_{sequential}}{T_{parallel}(N)} \tag{3.2}$$

The definition of speedup must, however, refer to a small gain sue because real problems cannot be fully implemented in the limited memory of a single processor, especially in distributed-memory systems (Fox *et al.*, 1988). This caveat suggests an important advantage of distributed-memory parallel computers: the ability to handle much larger problems than serial computers. An estimate of $T_{sequential}$ can be obtained by running a small problem on a serial machine and then extrapolating the result to the real problem.

The efficiency, ε, of a parallel algorithm running on N processors is defined as

$$\varepsilon = \frac{S(N)}{N} \tag{3.3}$$

Speedup, S, ranges from 1 to N and, consequently, efficiency is bounded by 0 and 1. Amdahl's law (Quinn, 1987, p. 19) states that, first, a small proportion of sequential operations can limit significantly the speedup achieved by a parallel algorithm and, second, that all processors must maintain some minimum level of efficiency or computing power will be wasted. As more processors are used, however, it becomes increasingly difficult to use them efficiently because, to maintain a constant level of efficiency, the fraction of computation that is serial must be proportional to the inverse of the number of processors (Quinn, 1987, p. 46).

In SIMD systems with fine grain size, parallelism may be implemented at several levels — normally that of individual instructions. With large grain sizes, however, each concurrent process actually is a sequential one. Algorithms with "sequential" components that affect their speedup are called loosely synchronous or asynchronous; such components include input and output, load balancing, and computation and communication overheads. Although some trade-offs are made among data distribution, message-passing, and computation to minimize total overheads, the ratio of these "sequential" components to parallel computation must be minimized in large grain size implementations. Asynchronous or loosely synchronous processing can, therefore, achieve a high speedup with a very large amount of parallcl computation. In an ideal case, the speedup of a parallel algorithm with N processors is $\Theta(N)$, which means that the speedup is equal to the number of processors used — called linear speedup. Arguments about the possibility of superlinear speedups are found in the literature (see Quinn, 1987, for example).

Effective sequential algorithms may not be suitable for parallelization (Fox *et al.*, 1988; Healey and Desa, 1989; Amstrong and Densham, 1992) and Amdahl's law can be used to determine their suitability (Quinn, 1987). Researchers who need to reorganize an algorithm or develop a new one for parallel processing must identify the best ways to parallelize it, co-ordinate its concurrent processes, balance processor workloads, and minimize overheads.

3.2.3 *Pipelining and parallelism*

High-performance parallel processing usually exhibits a great deal of concurrency that can be achieved by pipelining or parallelism. Pipelining divides a computation into a number of steps, each of which is executed by one of the pipeline's processors. The output of one step is the input to the next one, as in an assembly line. If all the steps require the same amount of processing time, then the multiplicative increase in throughput is equal to the number of steps in the pipeline when it is full. Throughput is degraded from this level, however, because it takes some time to fill and drain the pipe at the beginning and end of processing.

In parallelism, a computation is executed simultaneously on multiple pieces of data or a problem is solved concurrently by multiple processors. This is called partitioning or decomposition. Parallelism can be employed at different levels which depend on the parallel architecture used. SIMD systems usually exploit parallelism at the instruction level so that the inherently concurrent components in a computation, such as the elements of a loop, are executed in parallel. For MIMD architectures, a problem is usually divided into sub-problems that are solved by individual processors; by integrating the solutions to the sub-problems, the final solution is produced. There are three major approaches or paradigms for parallelism: event, geometric, and algorithmic parallelism (Healey and Desa, 1989).

Event or Task Farm Parallelism uses a master or control processor to distribute tasks to slave or worker processors during processing and then integrate the results. There are two major features of this approach: first, each slave processor runs the same code on its own specific dataset and, second, dynamic load balancing occurs because the master sends a new task to a worker when it becomes idle. Communication overhead during the dynamic control may degrade the performance of this approach, however.

Geometric or Data Parallelism partitions spatial domains into sub-regions or sub-sets. Individual partitions are handled by different processors and, for some problems, data must be passed from neighboring sub-region — such interregional communication degrades performance. The organization of processing and communication among processors is very important for this type of parallelism. A tightly synchronous model, for example, requires simultaneous processing on all processors and inappropriate parallelism may cause insufficient hardware utilization (Healey and Desa, 1989). While loosely synchronous processing with balanced workloads will achieve an expected level of performance, unbalanced workloads may cause unpredictable levels of inefficiency. Thus, in partitioning a problem, processing within a sub-region must be maximized and interregional communications minimized.

Algorithmic or Procedural Parallelism occurs when each processor works on a specific task and blocks of data pass through processors in a certain order — like a pipeline (Healey and Desa, 1989). This approach may lead to difficulties at the implementation stages because of communication and configuration problems. In general, data parallelism is better than procedural parallelism for most problems. When combined with other approaches, however, algorithmic parallelism is useful for special types of problems. Other, similar classifications of parallel paradigms can be found in Carriero and Gelenter (1990).

3.2.4 *Decomposition*

Parallelism requires that a problem is decomposed into concurrent sub-problems and the method of decomposition used is a crucial determinant of processing performance. Three general methods are used to decompose a problem into sub-parts: complete (perfectly parallel) decomposition, domain decomposition, and control decomposition (Ragsdale, 1991).

Complete decomposition divides a problem or application into a set of processes that require communication only to initialize individual sub-sessions and to collect results from them. The order of individual sub-processes is not important and no special synchronization is required during processing. A simple way to implement a complete decomposition is to execute equivalent sequential operations on all processors. Only when the number of partitions is greater than the number of processors is extra communication required to distribute tasks over processors. This method is similar to event parallelism and also to some special cases of data parallelism in which the data in each sub-problem are independent.

Domain decomposition divides a set of objects or data, that define the scope of a problem, into grains on each process or to achieve a high degree of concurrency (Fox *et al.*, 1988). Domain decomposition is well suited to the following three types of problems (Ragsdale, 1991):

- Problems with a static data structure. Raster-based image processing and cell-based grid modeling are both natural cases for domain decomposition.
- Problems with a dynamic data structure that are restricted to a single entity: searching for shortest paths over any type of network, for example.
- Problems with a fixed domain but dynamic computation. This type includes locating a number of facilities to meet demands for services in a given area.

Domain decomposition methods that correspond geometric or data parallelism can be used on most problems. While the domain's geometry and the mount of computation required significantly affect the decomposition (Fox *et al.*, 1988; Armstrong and Densham, 1992), the maximization of local operations and minimization of interprocessor communication are critical. Since domain decomposition often is used with distributed-memory MIMD computers, especially for spatial problems, other related issues are discussed in the following.

Control decomposition focuses on the flow of control in an application. This approach can be used when axed domain or domain decompositions are not appropriate (Ragsdale, 1991). There are two techniques for control decomposition: functional decomposition and manager/worker decomposition. In functional decomposition, a problem is viewed as a set of operations that define sequential computing processes. Each process may be carried out by one processor, and the physical relations between processes can be used to build a communications scenario. The granularity of the decomposition is determined by the number of processors in the system and the amount of computation associated with each function. In contrast, the manager/worker technique uses a manager process to farm out the tasks to each of the processors (workers) and to track progress as the workers report back with completed tasks. This dynamic allocation of computing tasks to processors makes more efficient use of computing resources for problems with variable domains. Functional decomposition is naturally suited to algorithmic parallelism and manager/worker decomposition corresponds well to event or task farm parallelism.

Another approach to decomposition views object-oriented programming (OOP) as a formalization of domain and/or control decomposition because both data and functions must be considered to create objects. Ragsdale (1991) described concurrent OOP and a detailed discussion can be found in Agha (1990). In many real-world applications, however, no single decomposition technique may be adequate for very large and complex problems. Different decomposition methods may be applied to each component or layer of a complex problem; this is called layered decomposition (Ragsdale, 1991) or composed decomposition.

3.2.5 *Communication and synchronization*

Although synchronous processing is automatic in SIMD systems, the coordination of comment processes on MIMD systems is a fundamental issue because communication often requires synchronization among processes. Communication is achieved by using global variables in shared-memory system and message passing in distributed-memory systems. Synchronization constrains the ordering of events in shared-memory systems, such as ordered reads from and writes to a shared memory; methods used to enforce synchronization include busy-waiting, semaphores, and monitors. In distributed-memory systems, however, synchronization controls interference and enables message passing to distribute data to, and collect results from, individual processes. Synchronous message passing is so named because a message sender delays processing until the corresponding receiver accepts the message, the message is passed, and both processors can resume their tasks. In contrast, buffered message passing uses buffers to store messages that have been sent but not yet received (Quinn, 1987).

3.2.6 *Load balancing*

A crucial issue for MIMD processing is how to assign tasks or datasets to processors so that processing time is minimized. If processors receive a variable number of processing tasks, those with a small number of tasks will finish their work earlier and, overall, there are wasted computing resources. If all processors have similar workloads, however, they will finish at about the same time. The load balancing problem can be stated as follows:

Given the total workload, T, and the number of processors, N, used in the system, assume that processor i will take a workload, t_i, such that

$$T = \sum_{i=1}^{N} t_i \tag{3.4}$$

Thus, the overall processing time required is defined by

$$T_{overall} = \max_{1 \le i \le N}(t_i) \tag{3.5}$$

Processing time is maximized when one processor is assigned the total workload

$$T_{overall} = T \tag{3.6}$$

and is minimized when all processors are assigned the same workloads

$$T_{overall} = \frac{T}{N} \tag{3.7}$$

The workload of simple, homogeneous problems can be divided into equal parts, each of which is assigned to a processor-static load balancing. For some complex and inhomogeneous problems, however, a dynamic load balancing method often is required (for a detailed discussion, see Fox *et al.*, 1988).

3.2.7 *Data dependencies*

Data dependencies are of two types: vertical and horizontal. Vertical dependency means that a process requires as an input the output of a previous process; it is caused by the interdependency of data items at the instruction or loop level — such as during data input and output in low level parallelism and when nested indexing is used for array elements (Wolfe and Banerjee, 1987). While there is little that can be done for input–output, because it is usually performed sequentially, careful design of nested loops and their allocation to processors can help in shared-memory or SIMD processing (Armstrong and Densham, 1992). Wolfe and Banerjee (1987) defined low-level data dependencies and discussed transformation methods for parallel processing. While vertical data dependency is normal in pipelined processing, it must be minimized in concurrent processing and often can be ignored in large grain size parallel processing because low-level data dependencies exist only within a single grain.

Horizontal data dependency occurs in partitioned and concurrent processes either when elements of the same array are accessed by multiple processors or when data are required from neighboring processes. In shared-memory systems, concurrent reading of array elements is allowed but only exclusive writing normally is permitted. The need to access data from neighboring processes arises in distributed-memory processing. The severity of such horizontal dependency largely is determined by the scope of the required operations or computations: it is most severe in process with global scope. When processes have limited scope, distributing expanded sub-datasets to each processor reduces or eliminates data dependency (Armstrong and Densham, 1992; Ding and Densham, 1994a). In sum, parallel algorithms with appropriate decomposition(s), well-balanced workloads, and optimized communication schemes will yield high levels of performance in many applications. Such issues are fundamental to the design, development, and implementation of new algorithms for parallel geoprocessing. To assist researchers in designing new algorithms, strategies for achieving these goals are required.

3.3 Parallelism in Spatial Modeling

3.3.1 *Spatial tessellations and sub-divisions*

One of the major features of space, the domain of spatial modeling, is its divisibility. Any region of interest can be divided into sub-regions or tiles, with each tile being linked to one or more points. A mosaic of tiles is called a tessellation. Regular tessellations are formed by rectangles, triangles, diamonds, hexagons, trapezoids, and parallelograms. Upton and Fingleton (1985) classified tessellations with straight-edged tiles into three special types: Dirichlet, Delaunay, and line tessellations. A Dirichlet tessellation (Voronoi or Thiessen polygons) divides space into a mosaic of convex polygons; every location within a polygon is closer to its focal point than to that of any other polygon. A Delaunay tessellation or triangulation is simply the dual of the Dirichlet tessellation. Finally, one line tessellation, the L-mosaic, is obtained by throwing lines at random onto the plane.

In addition to these special tessellations, regular tessellations, especially square or rectangular ones, have been widely used in GIS and spatial modeling. The most popular tessellation, the quadtree, recursively partitions space into a set of squares (Samet, 1990). Quadtrees adapt the binary search tree to two dimensions and have been widely used in spatial data analysis and GIS applications, such as two-dimensional run-encoding, the analysis and display of digital elevation data, and the determination of proximal polygons (Mark and Lauzon, 1984; Mark, 1987). Various two-dimensional orderings of space are used as quadtree codings; popular orderings — including Morton order, row order, row-prime order, and Pi-order — exhibit different spatial relationships and spatial properties (Goodchild and Grandfield, 1983; Goodchild and Mark, 1987; Abel and Mark, 1990). Although squares are the basic unit of division in quadtrees, quad-division can be extended to rectangular space and point sets. An adaptive recursive tessellation of rectangles, for example, can be used to divide a rectangular space to provide a more flexible tool for spatial data storage and analysis.

Spatial sub-division also has been used to tile geographical databases (Goodchild, 1989; Goodchild and Gopal, 1989). Furthermore, the concept of spatial recursive sub-division (SRS) has been extended to a global sphere; Dutton (1984) has proposed an SRS tiling scheme called the Quaternary Triangular Mesh (QTM). In this scheme, the spheroid is projected onto an octahedron aligned so that two vertices coincide with the poles, and the remaining four occur on the equator at longitudes of 0, 180, 90 West and 90 East. Each triangular facet is recursively sub-divided by connecting the midpoints of its sides.

3.3.2 *Data models and domains of spatial modeling*

Spatial models typically use a finite number of discrete, digital objects to capture the real geographic variation of a spatial problem. The rules in a data model define the objects used and their relations, how geographical variation is represented, and the set of processes and analyses that can be carried out. Goodchild (1991) identified five general types of spatial data analysis based on their underlying data models: (1) points, (2) spatial objects with attributes, (3) networks of links and node, (4) spatial interaction models, and (5) raster methods. The second and the fourth types use the same data model-a matrix representation.

Alternative data models form different domains for spatial modeling. Using data models from both object and field views of the world, domains for spatial modeling can be classified into six basic types (Table 3.1).

Table 3.1. A classification of domains for spatial modeling based on data models.

Domain types	Domain members	Examples of spatial modeling
Continuous space	Points, lines, areas	Dirichlet tessellation
Networks	Nodes and links	Shortest path problems
Cell-based grids	Cells or pixels	Hill shading on DEMs
Point sets	Points	Point pattern analysis
TIN	Individual triangles	Visibility analysis on TINs
Matrix	Matrix elements	Spatial interaction model

Source: Ding and Densham (1996).

- Continuous space corresponds to the data model in the object view. The domain can be viewed as a continuous and, if necessary, bounded area with a special reference system and distance metric. The objects in the space are points, lines, and areas.
- A network is a collection of points (nodes or vertices), some or all of which are connected by lines (links or edges). Links normally have attributes whereas nodes are used to store connectivity information among links; when they represent a special type of spatial object, however, nodes have their own attributes. Links and nodes are the domain elements and modeling is based on their attributes, topology, and geometry.
- A cell-based grid is a uniform, partitioned space consisting of pixels or cells. Each cell stores a numeric value that represents a continuous or discrete variable. A grid's resolution is determined by the size of its cells which, in turn, depends upon the data and type of analysis required. Cells or pixels are the spatial elements in this domain.
- A point set is a domain that consists only of points. Each point has its own attributes that describe its relation to other points.
- A triangulated irregular network (TIN) is a representation of a surface derived from irregularly spaced sample points and breakline features. Each sample point has a pair of co-ordinates (x, y) and a surface value (z); points are connected by edges to form a set of non-overlapping triangles, the basic elements of the domain.
- A matrix is in a spatial domain that reduces space to an array of spatial relations between pairs of objects. These relations often capture only an abstracted form of topological data because geometrical data often are lost when a matrix is compiled (Densham and Armstrong, 1993). The cells of the matrix are the domain elements.

3.3.3 *Classifications of spatial modeling for parallelism*

Any classification of spatial problems for parallel processing must focus on the pertinent characteristics of those problems, particularly their domains. Two characteristics of a domain are crucial in designing strategies for its decomposition: regularity and homogeneity. A domain's regularity determines how easily it can be partitioned geometrically into equal areas while the homogeneity of its member affects the balance among partition workloads. A regular domain can be partitioned using a simple geometrical scheme in which each partition has the same size sub-domain,

Table 3.2. A Classification of spatial problems based on their domains.

Domain features	Regular	Irregular
Homogenous	Type 1	Type 2
Inhomogeneous	Type 3	Type 4

Source: Ding *et al.* (1992).

for example, and all the members of a homogeneous domain are of the same type and require the same type of processing. A lack of homogeneity or regularity can increase the amount of inter-processor communication required, however, and lead to workload imbalances (Ding *et al.*, 1992). Of the six types of domains for spatial modeling discussed above, most are neither regular nor homogeneous for real problems. Such problems can be classified into four classes using as dimensions the regularity of their domains and the homogeneity of their domain members (Ding *et al.*, 1992) (Table 3.2).

- *Type 1.* Spatial problems with regular and homogeneous domains (Figure 3.l) have members of the same type and domains that can be decomposed using a simple geometrical approach. Thus, optimal domain decomposition and balanced loads can be achieved quite easily.
- *Type 2.* Spatial problems with irregular but homogeneous domains (Figure 3.2) have members of the same type but cannot easily be partitioned into equal parts. In this case, optimal domain decomposition and load balancing are possible and can be achieved by using static decomposition and extra work.
- *Type 3.* Spatial problems with regular but inhomogeneous domains (Figure 3.3). Although these problem domains can be partitioned easily into equal-sized parts, load balancing among the partitions is problematic because the members of the domain are of different types. Dynamic decomposition may be required to balance workloads.
- *Type 4.* Spatial problems with irregular and inhomogeneous domains (Figure 3.4) often are difficult to decompose into sub-parts with balanced workloads because they contain members of different types and have uneven geometry. A relatively balanced decomposition is possible using a dynamic decomposition technique. In some cases, other strategies may be required to complete the decomposition.

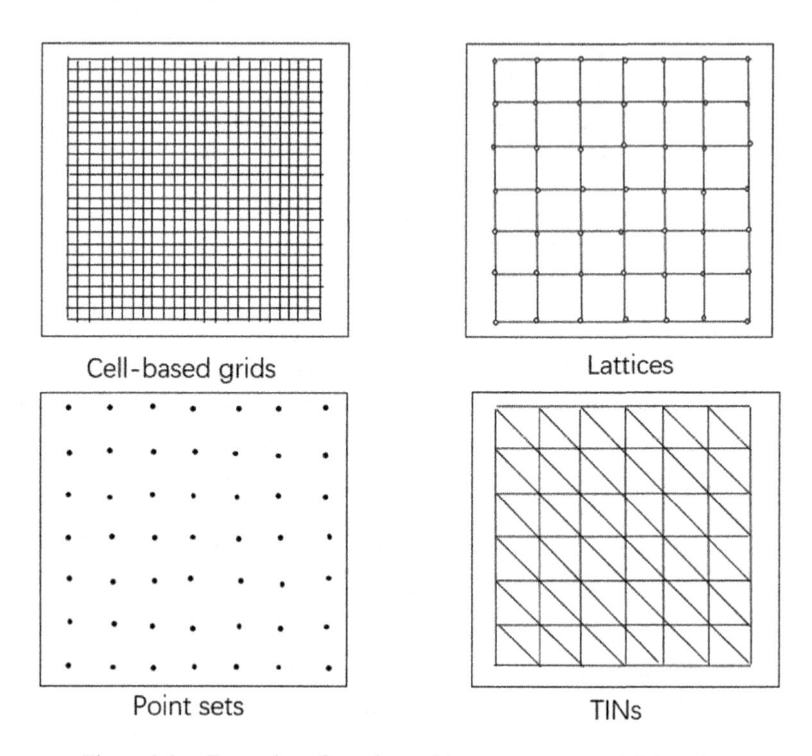

Figure 3.1. Examples of regular and homogeneous spatial domains.

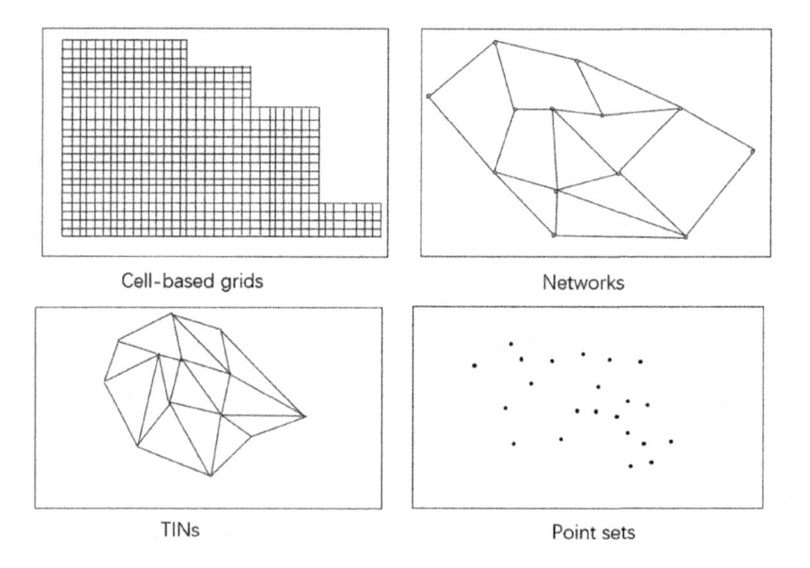

Figure 3.2. Examples of irregular and homogeneous spatial domains.

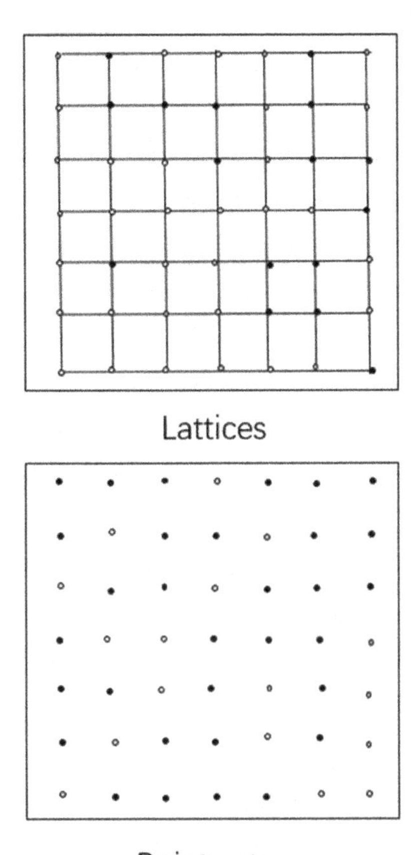

Figure 3.3. Examples of regular and inhomogeneous spatial domains.

Figures 3.1–3.4 illustrate the four categories of spatial domains for different data models. Cell-based grids are always homogeneous problems, because they have uniform cells, whereas networks can fall into all four categories of the classification. While a lattice is always a regular problem, the attributes of its vertices determine whether it is homogeneous or inhomogeneous; some of the vertices in location-allocation models, for example, represent candidate nodes and/or demand nodes, whereas others only depict street intersections. Similarly, TINs and point sets may belong to any one of the four categories. Although it is an

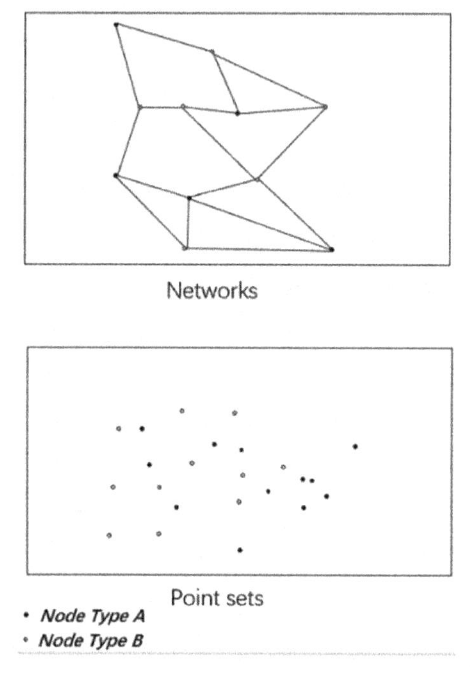

Figure 3.4. Examples of irregular and inhomogeneous spatial domains.

aspatial domain, a matrix is often thought of as a regular computational problem.

Interprocess communications are required to pass data between neighboring processes but often degrade the performance of parallel processing. In distributed-memory MIMD systems, interprocess communications usually are caused by horizontal data dependencies among the processes. The scope of spatial modeling determines the range of horizontal data dependencies: the larger the scope, the wider the range of data dependence. Consequently, different methods of parallelism are required to accommodate the effects of different degrees of scope and Tomlin's (1990) classification of grid operations provides the basis for a taxonomy of parallelism in spatial problems.

Spatial modeling with local scope: A model's final result consists of all sub-results from local operations. The domain can be decomposed without overlap because its range of horizontal data dependence is zero.

Spatial modeling with neighborhood scope: Solutions for this category of model, such as slope and aspect models on DEMs, synthesize the results from neighborhood operations. Such domains can be decomposed into overlapping neighborhoods.

Spatial modeling with regional scope: An anal result is assembled from sub-results for the overlapping sub-domains that completely cover the original domain. The sizes of the sub-domains are determined by the scope of operations and medium-range horizontal data dependencies.

Spatial modeling with global scope: All members of the spatial domain are used to generate a final result. Any missing or modified members change the final result; consequently, results from sub-domains are difficult to assemble into an anal result. Such models exhibit a large range of horizontal data dependence.

3.4 Strategies for Parallelizing Spatial Modeling

The identification of problem structures is very important for the design of parallel algorithms. Fox (1991) presented a classification of problem structures, and their relation to parallel architectures, that identifies three broad types of problem structures based on the differences in both spatial (data) and temporal (control) aspects (Table 3.3).

Synchronous problems are data parallel with the restriction of the same control: each data element is controlled by the same algorithm. Such problems are regular in both space and time and are naturally suited to SIMD architectures. Slope and aspect modeling on digital elevation

Table 3.3. A classification of problem structures.

Classes	Spatial/ data aspect	Temporal/ control aspect	Natural support hardware
Class I: Synchronous	Regular	Regular	SIMD
Class II: Loosely synchronous	Irregular	Regular	Distributed-memory MIMD
Class III: Asynchronous	Irregular	Irregular	MIMD

Source: Fox (1991).

models are typical examples of this category. Loosely synchronous problems are irregular in their spatial (data) aspects but regular in their temporal (control) aspects. This type of problem typically is data parallel with control by distinct algorithms and, because the spatial structure often changes dynamically, adaptive algorithms normally are required (Fox, 1991). Massively parallel, distributed-memory MIMD machines with asynchronous hardware architectures are most suitable for such problems. Asynchronous problems are irregular in both space and time, often exhibit functional or process parallelism, and usually are hard to parallelize. Fortunately, the sub-division property of spatial domains provides a way to parallelize such problems. Parallel algorithms for each of these classes of spatial problems can be developed using decompositions of spatial domains and spatial models that exploit only MIMD, distributed-memory architectures. This is possible because, in addition to their asynchronous processing capabilities, such architectures can simulate synchronous processing.

3.4.1 *Simple spatial partitioning for spatial domain decomposition*

Domain decomposition of spatial problems must take advantage of the sub-division property of space. Only square or rectangular tessellations of space are implemented easily in computers because of the nature of spatial co-ordinate systems. A rectangular tessellation of a regular space can be generated by a one- or two-dimensional partitioning and each has different purposes and effects. Regular space can be split easily into any number of equal parts, odd or even, by one-dimensional partitioning. Two-dimensional regular division always creates an even number of equal parts. Spatial data are easier to divide, distribute, and assemble over one dimension than two. A cell-based grid, for example, can be split and merged using either rows or columns. The big difference between using one or two dimensions is the length of the edges in the partitions: it is often less for a two-dimensional sub-division than for a one-dimensional partition (Fox *et al.*, 1988; Ding and Weber, 1993). The lengths of the edges usually determine the amount of communication among neighbors and, consequently, affect the performance of parallel processing. Thus, a two-dimensional partitioning often is more efficient than a one-dimensional one because less communication is required.

Other types of regular tessellations, based on a parallelogram, rhombus, or triangle, are sometimes useful for decomposing spatial domains. Transformations or projections of the co-ordinate system can be used to convert these special regular tessellations into rectangular ones. For some directional spatial modeling, such as the detection of cast shadows along the axis of illumination on a DEM, an appropriate decomposition is one-dimensional partitioning perpendicular to the axis of illumination (Ding and Densham, 1994a). If the light direction is not parallel to one of the sides of a rectangular space, a regular partitioning based on intervals of equal distance will result in unequal workloads. In this case, an equal-area partitioning scheme can be used to balance workloads in a homogeneous space (Figure 3.5).

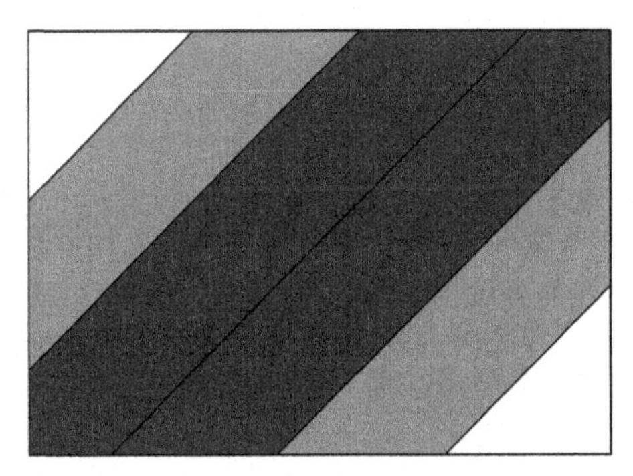

Directional Equal Distance Partitioning

Directional Equal Area Partitioning

Figure 3.5. Directional irregular partitioning.

Regular division of irregular or inhomogeneous spatial domains causes unbalanced workloads. Thus, spatial partitioning must be adjusted to divide equally spatial objects or elements that are irregularly distributed. Each spatial object has a location described by a reference system in the space. Partitioning of irregular or inhomogeneous space proceeds by sorting or ordering objects in space and then dividing them into equal groups (Figure 3.6).

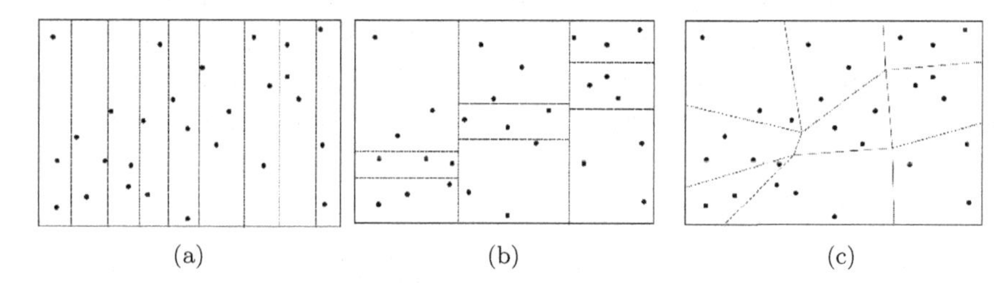

Figure 3.6. Spatially adaptive partitioning.

Experiments show that a very efficient way to partition an inhomogeneous space into balanced parts is to use an alternating net (Hinz, 1990). A net of equally sized rectangles is placed over the regular space and then distorted by moving as many net points as necessary to equalize the partition workloads while maintaining the topology (Figure 3.6(c)).

3.4.2 *Recursive spatial sub-divisions for spatial domain decomposition*

Quadtrees are not only applied to spatial data structures and spatial databases but also to sequential algorithm designs and parallel algorithm development (Bestul, 1990).

Quadtree-based recursive sub-divisions generate square-shaped partitions of a square space (Figure 3.7(a)). The extension of this well-known idea to rectangular spaces has been proposed by several researchers. Ding and Weber (1993), for example, designed and implemented a rectangle-based recursive quad-division approach for a sequential algorithm to improve the efficiency of spatial search (Figure 3.7(b)). Tsui and Brimicombe (1992) use a variable ratio of decomposition at each level in the tree (Figure 3.7(c)); this adaptive recursive tessellation (ART) has been used to represent hierarchical spatial data and maps. All these methods can be used to partition a space if it is regular, homogeneous, and can

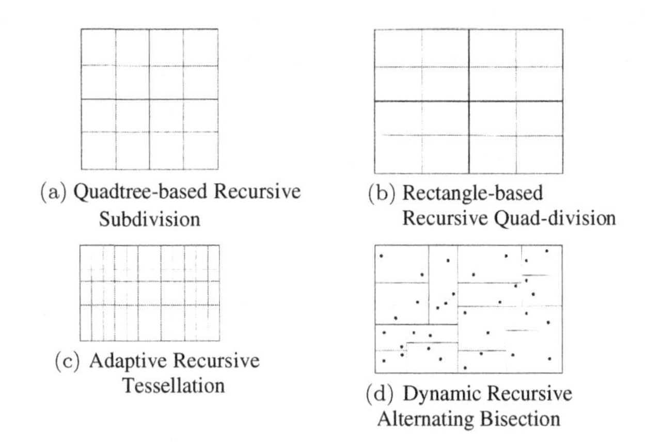

(a) Quadtree-based Recursive Subdivision

(b) Rectangle-based Recursive Quad-division

(c) Adaptive Recursive Tessellation

(d) Dynamic Recursive Alternating Bisection

Figure 3.7. Spatially recursive sub-divisions.

be decomposed into uniform cells. For irregular or inhomogeneous spatial domains, however, partitioning by quadtree or rectangles yields unbalanced workloads. To solve this problem, a dynamic recursive alternating bisection (DRAB) method can be used to partition unevenly distributed spatial objects or elements. First, a straight line is used to divide a set of spatial objects or points into two equal partitions; subsequently, each partition is divided recursively into two equal sub-partitions by a straight line perpendicular to the one used previously. This hierarchical alternating bisection forms a two-dimensional partitioning over an irregular or inhomogeneous space (Figure 3.7(d)), creating well-balanced partitions (Belkhale and Banerjee, 1990; Hinz, 1990; Ding, 1993).

Table 3.4 lists several types of recursive spatial partitioning used for domain decomposition. If a quadtree or binary tree structure is used to organize the processors, the operations implementing recursive partitioning can be parallelized and executed concurrently. Thus, instead of using only the master, multiple processors decompose and distribute data.

3.4.2.1 *Spatial overlapping versus non-overlapping decompositions*

Spatial domain decomposition, using either single-level partitioning or recursive sub-division, divides a domain into a number of small parts that usually do not overlap. For spatial models of local scope, a non-overlapping decomposition is the most efficient way to parallel the problem. Results from the sub-domains are assembled to yield a final result.

Table 3.4. Types of spatial recursive partitioning.

Methods	Basic units	Number of partitions in each level	Suitable domain
Quadtree	Square	4	Homogeneous and square domains
Rectangle-based quad-division	Rectangle	4	Homogeneous domains
Adaptive recursive tessellation	Rectangle	≥4	Homogeneous domains
Dynamic recursive alternating bisection	Rectangle	2	Inhomogeneous domains

Source: Ding and Densham (1996).

Other types of geoprocessing have different ranges of horizontal data dependency and using non-overlapping sub-domains without neighbor-to-neighbor communication may lead to sub-results that cannot be combined and, hence, incorrect results. Performance is significantly degraded by heavy communication, however, because it must be synchronized. One solution is to extend each sub-domain by the range of horizontal data dependence (Figure 3.8), yielding an overlapping domain decomposition. There is always a trade-off between the amount of communication and the data overhead that results from overlapping sub-domains.

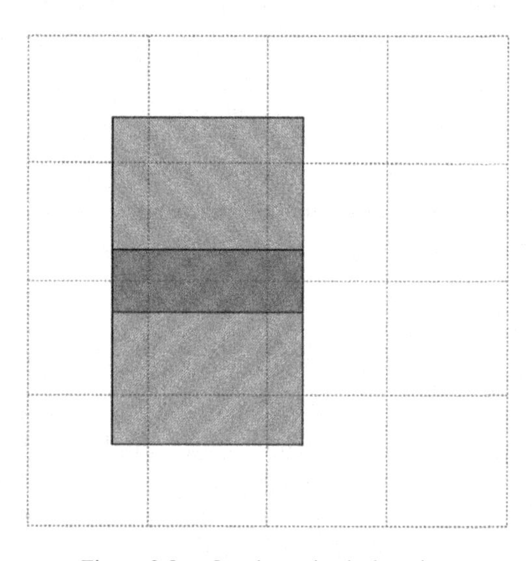

Figure 3.8. Overlapped sub-domains.

Overlapping domain decomposition is a better way to parallelize modeling with neighborhood scope than non-overlapping decomposition because of the restricted range of data dependence. The small amount of data overhead incurred obviates heavy communication and its associated synchronization and also has the benefit of increasing the grain size. For problems of global scope, however, overlapping domain decompositions are inefficient or ineffective because they greatly increase data overhead. In such cases, other decomposition methods usually are required, such as process decomposition or functional decomposition.

3.4.2.2 *Spatial strategies for different types of spatial modeling*

Table 3.5 outlines spatial strategies that can be applied to different types of problems. Spatial models of local scope with regular and homogeneous domains can be parallelized using a simple spatial partitioning or recursive regular sub-division. One-dimensional partitions can be the easiest to use because they do not need neighbor-to-neighbor communication. Spatial models of neighborhood or regional scope, with regular and homogeneous domains, can be parallelized using two-dimensional overlapping partitions or recursive overlapping quad-divisions — the degree of overlap is determined by the range of the data dependency.

Models with irregular and homogeneous domains, and operations of local scope, should use non-overlapping, equal-area spatial partitioning to balance workloads. Neighbor-to-neighbor communication is not an issue because there is no horizontal data dependence. When scope is of a neighborhood or regional scale, alternative strategies can be applied. The first uses overlapping spatial partitions of equal area. In the second approach, the irregular and homogeneous domain is treated as if it was a regular and inhomogeneous one that can be parallelized using two-dimensional, overlapping adaptive partitioning or overlapping DRAB (Figure 3.9). With local scope, there is no need for overlap in the decomposition. Irregular and inhomogeneous domains also can be treated as regular and inhomogeneous spaces and overlapping or non-overlapping dynamic decomposition methods can be used to parallelize this type of spatial model.

Several approaches are available to parallelize spatial models of global scope. One method is to use non-overlapping strategies to partition computational units while keeping the original dataset undivided. Each processor is assigned one partition of the computational task and operates

Table 3.5. Spatial strategies for parallelism of spatial modeling.

Types of problems	Local scope spatial modeling	Neighborhood scope spatial modeling	Regional scope spatial modeling	Global scope spatial modeling
Regular and homogeneous domains	Non-overlapping one-dimensional partitioning Non-overlapping two-dimensional partitioning Non-overlapping recursive regular sub-divisions	Overlapping two-dimensional partitioning or recursive quad-division	Overlapping two-dimensional partitioning or recursive quad-division	Use non-overlapping strategies to decompose the domain members but keep the dataset undivided
Irregular and homogeneous domains	Non-overlapping equal area spatial partitioning Non-overlapping dynamic recursive alternating bisection	Overlapping equal area spatial partitioning Two-dimensional overlapping adaptive partitioning Overlapping dynamic recursive alternating bisection	Overlapping equal area spatial partitioning Two-dimensional overlapping adaptive partitioning Overlapping dynamic recursive alternating bisection	Use non-overlapping strategies to decompose the domain members but keep the dataset undivided

Regular and inhomogeneous domains	Two-dimensional non-overlapping adaptive partitioning Non-overlapping dynamic recursive alternating bisection	Two-dimensional overlapping adaptive partitioning Overlapping dynamic recursive alternating bisection	Two-dimensional overlapping adaptive partitioning Overlapping dynamic recursive alternating bisection	Use non-overlapping strategies to decompose the domain members but keep the dataset undivided
Irregular and inhomogeneous domains	Two-dimensional non-overlapping adaptive partitioning Non-overlapping dynamic recursive alternating bisection	Two-dimensional overlapping adaptive partitioning Overlapping dynamic recursive alternating bisection	Two-dimensional overlapping adaptive partitioning Overlapping dynamic recursive alternating bisection	Use non-overlapping strategies to decompose the domain members but keep the dataset undivided

Source: Ding and Densham (1996)

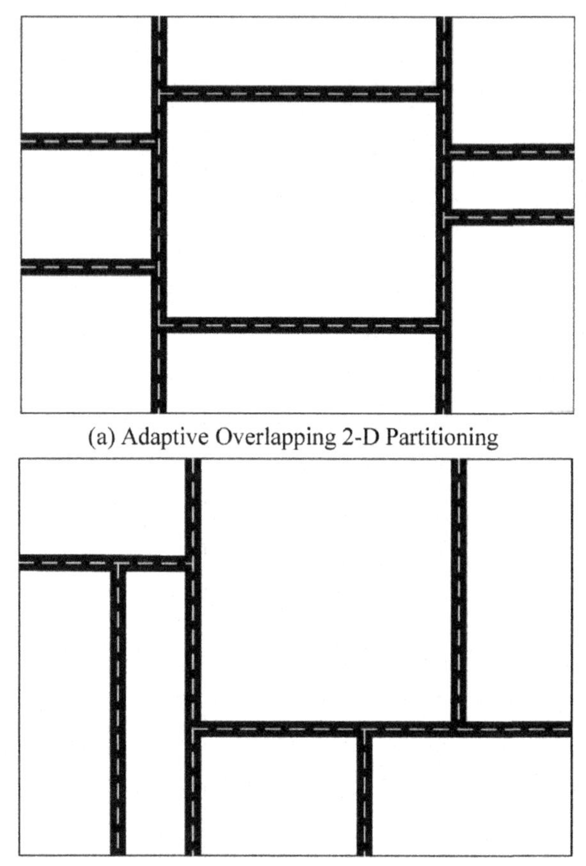

Figure 3.9. Dynamic overlapping partitioning.

on the entire dataset. Parallel processing using partitioned computational units and undivided datasets causes data overhead problems, however, which may degrade performance. Furthermore, the localization of global scope modeling through spatial domain decomposition is very expensive in terms of processing time. Consequently, it may be beneficial to use aspatial rather than spatial strategies of decomposition.

3.4.2.3 *Aspatial strategies for parallelism*

Several aspatial strategies can be used to parallelize global scope processing:

- Divide computational units into a number of groups and assign each to a processor with a pertinent dataset. Such groupings do not account for the spatial relations among them, which may be exploited to reduce the amount of computing required.
- Divide the entire process into a number of sub-processes and run in a pipeline fashion. This is a control decomposition.
- Apply the compute-aggregate-broadcast paradigm to ensure that all sub-results from individual processes can be assembled into a final result.

Although these approaches are suggested for global scope modeling, they are also suitable for other types of modeling. A combination of several strategies may be required for particular complex domains, for example, that have more restricted degrees of scope.

3.4.3 *Implementations of the strategies*

To demonstrate how the strategies described above are used to implement spatial modeling, two models are used: hill shading of a DEM and triangulation in raster space. Both models have been implemented in parallel on a MIMD architecture computer: a Transputer array that consists of four processors and their local RAM mounted on a PC add-in card. The Transputers can exploit the mass storage, graphical display, and input and output capabilities of the microcomputer host.

3.4.3.1 *A homogeneous spatial problem — Parallel hill shading on DEMs*

The spatial domain of hill shading on a distal elevation model (DEM) is the DEM itself. Although each cell in a DEM may have a different elevation value, all cells, except for the edge cells which are usually excluded from computational units, are treated as the same type of element in the hill shading model. Thus, a DEM can be viewed as a regular and homogeneous spatial domain. In most cases, a DEM with an irregular boundary simply can be converted into a regular and homogeneous domain. The irregular outline of the original domain can then be marked out on the new version.

The hill shading model has three components (Ding, 1992): self-shadows, S_{self}, modeled by a bi-directional reflectance distribution function with a fixed vertical viewer and a mobile light source; cast shadows, $S_{casting}$, detected by a rotation-transformation approach; and atmospheric scattering effects, $S_{scattering}$, simulated by a ball model. The model is

$$R(x, y) = S_{self}(x, y, el, az) * S_{casting}(x, y, el, az) + * S_{scattering}(x, y) \quad (3.8)$$

where (χ, y) is the location of the cell; *el* is the elevation angle of the light source; *az* is the azimuth of the light source; and γ is the ratio of atmospheric scattering over direct radiance.

Self-shadows are determined by the relative reflectance of pixels to the light Source:

$$S_{self}(x, y, el, az) = \text{Positive}(\sin(el) - \cos(el)*[XG_{xy}\sin(az) + YG_{xy}*\cos(az)]) \quad (3.9)$$

where

$$\text{Positive(a)} = \begin{cases} a, & \text{when } a > 0; \\ b, & \text{when } a \Leftarrow 0. \end{cases}$$

$$XG_{xy} = \{(z[x+1, y+1] + 2 * z\,[x, y+1] + z[x-1, y+1]) - (z[x+1, y-1] + 2 * z[x, y-1] + z[x-1, y-1])\}/(8 * d_x),$$

the gradient in the X direction (north–south);

$$YG_{xy} = \{(z[x+1, y+1] + 2 * z[x+1, y] + z[x+1, y-1]) - (z[x-1, y+1] + 2 * z[x-1, y] + z[x-1, y-1])\}/(8 * d_y),$$

the gradient in the Y direction (East–West); $z[x, y]$ is the elevation of a pixel (x, y); and d_x and d_y are the sin of a pixel. Atmospheric scattering effects are estimated by a ball model. The solid angle of the cell open to the sky is defined by a pixel's eight neighboring pixels and the relative atmospheric scattering effects are described as follows (Ding, 1992):

$$S_{scattering}(x, y) = \frac{1}{2} - \sum_{k=1}^{8} \frac{\arctan\left(\dfrac{z_k - z_{x,y}}{d_k}\right)}{8\pi} \quad (3.10)$$

where $z_{x,y}$ is the elevation of the target pixel (x, y); z_k is the elevation of the kth pixel in the 8 neighbors; and

$$d_k = \begin{cases} \text{the pixel size,} & \text{for the 4 neighbors;} \\ \sqrt{2} * \text{the pixel size,} & \text{for the 4 diagonal neighbors.} \end{cases}$$

Both the self-shadow and atmospheric scattering models are pixel-based, window operations (computed using a cell's eight neighbors) that are of neighborhood scope on a regular and homogeneous spatial domain. An overlapping, one-dimensional partitioning is used to parallelize the models (Ding and Densham, 1994a).

Cast shadows are detected by checking which surface cells are invisible from a light source that is assumed to be infinitely distant. The detection model uses a co-ordinate transformation method followed by a simple but full scan of the transformed elevations along the azimuth of the light source (Ding, 1992). The model is as follows:

$$S_{casting}(x,y) = \begin{cases} 0 \text{ (within cast shadows),} & \text{if } Z_T(x,y) \Leftarrow Z_{T_{previous\_max}}; \\ 1 \text{ (not in cast shadows),} & \text{otherwise} \end{cases} \tag{3.11}$$

where

$$Z_T(x, y) = z(x, y) * cos(el) + [x * cos(az) - y * sin(az)] * sin(el)$$

In contrast to the other components, the cast shadow model is a directional operation of global scope. If the direction of the light source is not parallel to either edge of the region (typically it is not), then a regular, one-dimensional, directional partitioning will cause unbalanced workloads. TMs irregular but homogeneous spatial problem requires a special decomposition method: directional and equal-area partitioning.

The parallel algorithm for this homogeneous and regular/irregular spatial problem has been implemented on a MIMD computer — a Transputer array of four processors (Ding and Densham, 1994a). The master processor reads the DEM data, partitions it, distributes the sub-sets to other worker processors, computes the existence and extent of cast shadows, and partitions these cast shadows and distributes them to worker processors, and then receives the results bask and assembles them into the anal results. After they receive partitions of the dataset from the master, the worker processors run the self-shadow and atmospheric scattering models on these overlapping partitions while the master is generating the cast shadows; the workers then receive the cast shadows from the master and integrate them with the locally computed self-shadows and atmospheric scattering effects before passing the results to the master. This algorithm runs different programs on different datasets at the same time, taking advantages of the asynchronous processing capabilities of MIMD computers. Solution times for different datasets are listed in Table 3.6.

Table 3.6. Solution times for parallel hill shading of different datasets.

Sites	Keating summit, PA	West Point, NY	Crater Lake, OR
Dataset size	128 × 128	256 × 256	256 × 256
Azimuth (°)	90	90	270
Elevation angler (°)	15	15	30
Serial processing (sec)	14.090	99.738	98.743
Parallel processing with four processors (sec)	6.088	31.076	29.602
Speedup	2.31	3.21	3.34
Efficiency (%)	58	80.2	83.4

Source: Ding and Densham (1996).

These results not only show the efficiency of the parallel algorithm but also demonstrate the effects of increasing grain size for a given number of processors: the larger the grain size, the greater the efficiency of the parallel algorithm. This occurs because as the grain size increases for each processor, the ratio of communication to computation decreases and each processor spends relatively more of its time computing than it does communicating. The highest efficiency achieved is 83.4% with a speedup of 3.34 for the four processors. A detailed discussion of this implementation is to be found in Ding and Densham (1994a).

3.4.3.2 *A regular and inhomogeneous spatial problem — Parallel triangulation in raster space*

Delaunay triangulation yields a spatial tessellation that is a useful data structure for GIS applications. Two different approaches are used to build the triangulation from a set of points in a two-dimensional space: direct creation from a set of points using the smallest co-circular criterion (Lee and Schachter, 1980) and conversion from a dual set of Thiessen polygons because the perpendicular bisectors of the facets of the Thiessen polygons form the Delaunay triangulation. Although the most popular algorithms for constructing both the tessellation and the triangulation are vector-based (Aurenhammer, 1991), raster-based algorithms provide different insights into the problem (Mark, 1987). The space containing a set of points or centroids is overlaid with a fine grid and each grid cell is allocated to its closest centroid-forming Thiessen polygons. A Delaunay triangulation can be derived by linking the centroids of the adjacent polygons. Mark (1987) has proposed refinements to the raster allocation method, including the use of a recursive quadtree to avoid "brute force" allocation of each cell. Ding and Weber (1993) extend this work using a rectangle-based, recursive quad-division of the space. This approach forms the basis of a parallel algorithm.

To localize the allocation of grid cells to their closest points, and thereby reduce computation significantly, both the space and its associated point set must be sub-divided. Recursive quad-division of both the grid and point set into sub-regions turns the grid into an inhomogeneous domain, however, because the points may be spread unevenly over the grid space. Thus, the amount of computation required for each cell differs across the sub-regions and, to balance processor workloads, the decomposition must account for this inhomogeneity.

The parallel algorithm consists of three steps: parallel preprocessing that divides the domain for multiple processors, concurrent processing that builds the local triangulations simultaneously, and parallel post-processing that merges the triangulations (Ding and Densham, 1994b). During parallel preprocessing, a dynamic, recursive, and alternating bisection (DRAB) is used to decompose both the grid space and the point set into overlapping sub-sets. DRAB is designed explicitly to account for the distribution of points. The grid space is first divided, in a direction perpendicular to one of its axes, into two sub-domains each with the same number of points. Each sub-domain and its overlapping point subset are then simultaneously and recursively sub-divided in alternating perpendicular directions. This parallel pre-process forms a binary tree in which each node represents a sub-division that is assisted by a single processor. During concurrent processing, each processor builds its local triangulation using the raster-based, recursive quad-division algorithm (Ding and Weber, 1993). This algorithm further divides the points and grid space into sub-sets until an efficient level is reached. A gamma-shaped local operator then identifies adjacency relations among points so that the local triangulations can be formed. To combine these triangulations during the parallel post-processing step, the binary tree formed during decomposition is used. Thus, the triangulations are assembled using a parallel double-index quick sort and merge approach. The details of the design and implementation can be found in Ding and Densham (1994b).

A serial version of the algorithm was implemented to measure the performance of the parallel algorithm. The only difference between the parallel and serial algorithms used is the way they divide space; while the parallel algorithm uses the DRAB approach, the serial version is built around rectangle-based quad-division. Since the DRAB method accounts for spatial relations and spatial distributions, the parallel algorithm is expected to be more efficient than the serial algorithm. Both parallel and serial algorithms have been tested using a number of randomly generated point sets. The point co-ordinates were converted to range from 0 to 400 units on both axes while the number of points in the datasets ranges from 10 to 450.

References

Abel, D. J. and Mark, D. M. (1990). A comparative analysis of some two-dimensional orderings. *International Journal of Geographical Information Systems*, **4**(1): 21–31.

Agha, G. (1990). Concurrent object-oriented programming. *Communications of the ACM*, **33**(9): 125–141.

Armstrong, M. P. and Rokos, D.-K. (1992). *Parallel terrain feature extraction.* University of Iowa.

Armstrong, M. P. and Densham, P. J. (1992). Domain decomposition for parallel processing of spatial problems. *Computers, Environment and Urban Systems*, **16**(6): 497–513.

Aurenhammer, F. (1991). Voronoi diagrams — A survey of a fundamental geometric data structure. *ACM Computing Surveys (CSUR)*, **23**(3): 345–405.

Belkhale, K. P. and Banerjee, P. (1990). A parallel algorithm for hierarchical circuit extraction. In ICCAD, pp. 236–239.

Bertsekas, D. P. and Tsitsiklis, J. N. (1989). A survey of some aspects of parallel and distributed iterative algorithms.

Bestul, T. H. (1990). Gower's mirror De L'Omme and the meditative tradition. *Mediaevalia*, **16**: 307–328.

Bestul, T. (1991). A general technique for creating SIMD algorithms on parallel pointer-based. In *Auto Carto 10: 10. International Symposium on Computer-Assisted Cartography*, Baltimore, Maryland, March 25–28, 1991; *ACSM 51st Annual Convention; ASPRS 56th Annual Convention*, Vol. 6, p. 428. *American Congress on Surveying and Mapping.*

Carriero, N. and Gelernter, D. (1990). *How to Write Parallel Programs: A First Course*. Cambridge: MIT Press.

Chaudhary, V. and Aggarwal, J. K. (1990). Generalized mapping of parallel algorithms onto parallel architectures. In *ICPP*, (2), pp. 137–141.

Collins, G. (1989). A fast recursive parallel algorithm for triangulation of unordered DTM's. *Cartography*, **18**(1): 21–25.

Dangermond, J. and Morehouse, S. (1987). Trends in hardware for geographic information systems. In *Proceedings of the 8th International Symposium on Computer-Assisted Cartography (Auto-Carto 8)*, Baltimore, Maryland, March.

Denning, P. J. and Tichy, W. F. (1990). Highly parallel computation. *Science*, **250**(4985): 1217–1222.

Densham, P. J. and Armstrong, M. P. (1993). Supporting visual interactive locational analysis using multiple abstracted topological structures. In *Autocarto-Conference-ASPRS American Society for Photogrammetry and Remote Sensing*, pp. 12–12.

DeWitt, D. and Gray, J. (1992). Parallel database systems: The future of high performance database systems. *Communications of the ACM*, **35**(6): 85–98.

Ding, Y. and Densham, P. J. (1994). A loosely synchronous, parallel algorithm for hill shading digital elevation models. *Cartography and Geographic Information Systems*, **21**(1): 5–14.

Ding, T. and Zheng, J. (1993). Implementation of destination-address block-location using an SIMD machine. In *Visual Communications and Image Processing'93*, Vol. 2094, pp. 1042–1053. SPIE.

Ding, Y. and Densham, P. J. (1994). A loosely synchronous, parallel algorithm for hill shading digital elevation models. *Cartography and Geographic Information Systems*, **21**(1): 5–14.

Ding, Y. and Densham, P. J. (1996). Spatial strategies for parallel spatial modelling. *International Journal of Geographical Information Systems*, **10**(6): 669–698.

Dutton, G. (1996). Encoding and handling geospatial data with hierarchical triangular meshes. In *Proceeding of 7th International Symposium on Spatial Data Handling*, Vol. 43. Netherlands: Taylor & Francis.

Franklin, S. B., Gibson, D. J., Robertson, P. A., Pohlmann, J. T., and Fralish, J. S. (1995). Parallel analysis: A method for determining significant principal components. *Journal of Vegetation Science*, **6**(1): 99–106.

Jones, D. (1999). *Example of Reference Style in Stylesheets: Format Using Ctrl+Shift+F8*. Singapore: World Scientific,

Goodchild, M. F. (1989). Modeling error in objects and fields. *Accuracy of Spatial Databases*, **113**.

Goodchild, M. F. (1991). Geographic information systems. *Progress in Human Geography,* **15**(2): 194–200.

Goodchild, M. F. and Grandfield, A. W. (1983). Optimizing raster storage: An examination of four alternatives. *Proceedings of Auto-Carto*, **6**(1): 400–407.

Goodchild, M. F. and Mark, D. M. (1987). The fractal nature of geographic phenomena. *Annals of the Association of American Geographers*, **77**(2): 265–278.

Goodchild, M. F. and Gopal, S. (Eds.). (1989). *The Accuracy of Spatial Databases.*

Harris, J. D. and Connell, H. E. T. (1985). An interconnection scheme for a tightly coupled massively parallel computer network. In *Proceedings of ICCD*, Port Chester, NY.

Healey, R. G. and Desa, G. B. (1989). Transputer based parallel processing for GIS analysis: Problems and potentialities. In *Proceedings of Auto-Carto 9 (Bethesda: American Congress for Surveying and Mapping)*, pp. 90–99.

Healey, R. G. and Desa, G. B. (1989). Transputer based parallel processing for GIS analysis: Problems and potentialities. *Proceedings of Auto-Carto*, **9**: 90–99

Hopkins, S., Healey, R.G., and Waugh, T.C. (1992). Algorithm scalability for line intersection detection in parallel polygon overlay. *Proc. 5th International Symposium on Spatial Data Handling,* IGU commission of GIS, Charleston, South Carolina, pp. 210–218.

Hopkins, S. and Healey, R. G. (1990). A parallel implementation of Franklin's uniform grid technique for line intersection detection on a large transputer array. Regional Research Laboratory for Scotland, Edinburgh, pp. 95–104.

Langran, G. (1991). Generalization and parallel computation. *Map Generalization: Making Rules for Knowledge Representation*, p. 204216. Longman: London.

Lee, D.-T. and Schachter, B. J. (1980). Two algorithms for constructing a Delaunay triangulation. *International Journal of Computer & Information Sciences*, **9**(3): 219–242.

Li, Z. (1992). Array privatization for parallel execution of loops. In *Proceedings of the 6th International Conference on Supercomputing*, pp. 313–322.

Mark, D. M. (1987). Recursive algorithm for determination of proximal (Thiessen) polygons in any metric space. *Geographical Analysis*, **19**(3): 264–272.

Mark, D. M. and Lauzon, J. P. (1984). Linear quadtrees for geographic information systems. In *Proceedings of the International Symposium on Spatial Data Handling*, Vol. 2, pp. 412–430. Zurich.

Modi, J. J. (1988). *Parallel Algorithms and Matrix Computation.* Oxford: Clarendon.

Mower, J. E. (1992). Building a GIS for parallel computing environments. In *Proceedings of the 5th International Symposium on Spatial Data Handling*, pp. 219–229.

Peucker, T. K. and Douglas, D. H. (1975). Detection of surface-specific points by local parallel processing of discrete terrain elevation data. *Computer Graphics and Image Processing*, **4**(4): 375–387.

Preston, K. (Ed.). (1982). *Multicomputers and Image Processing: Algorithms and Programs*: Amsterdam: Elsevier.

Puppo, E., Davis, L., De Menthon, D., and Teng, Y. A. (1994). Parallel terrain triangulation. *International Journal of Geographical Information Science*, **8**(2): 105–128.

Quinn, M. J. (1987). *Designing Efficient Algorithms for Parallel Computers.* New York: McGraw-Hill, Inc.

Quinn, M. J. and Deo, N. (1984). Parallel graph algorithms. *ACM Computing Surveys (CSUR)*, **16**(3): 319–348.

Ragsdale, S. (Ed.). (1991). *Parallel Programming*. New York: McGraw-Hill, Inc.

Samet, Hanan. (1990). *The Design and Analysis of Spatial Data Structures*, Vol. 85. Reading, MA: Addison-Wesley.

Sandu, J. S. and Marble, D. F. (1988). An investigation into the utility of the Cray X-MP supercomputer for handling spatial data. In *Proceedings of the 3rd International Symposium on Spatial Data Handling*, pp. 253–266. Australia: Sydney.

Smith, L. (1999). *The Next Sample Reference.* London: Imperial College Press.

Smith, C. G., Pepper, M., Newbury, R., Ahmed, H., Hasko, D. G., Peacock, D. C., Frost, J. E. F., Ritchie, D. A., Jones, G. A. C., and Hill, G. (1989). One-dimensional quantised ballistic resistors in parallel configuration. *Journal of Physics: Condensed Matter*, **1**(37): 6763.

Tsui, P. H. Y. and Brimicombe, A. J. (1997). Adaptive recursive tessellations (ART) for geographical information systems. *International Journal of Geographical Information Science*, **11**(3): 247–263.

Upton, G. J. G. and Fingleton, B. (1985). *Spatial Data Analysis by Example: Categorical and Directional Data.* Hoboken: Wiley.

Verts, W. T. and Thomson, C. L. (1988). Parallel architectures for geographic information systems. In *Technical Papers of the ACSM-ASPRS Annual Convention (Bethesda: American Congress on Surveying and Mapping)*, pp. 101–107.

Ware, A. and Kidner, D. (1991). Parallel implementation of the Delaunay triangulation within a transputer environment. In *Proceedings of the 2nd European Conference on Geographical Information Systems*, Vol. 2, pp. 1199–1208.

Wolfe, M. and Banerjee, U. (1987). Data dependence and its application to parallel processing. *International Journal of Parallel Programming*, **16**(2): 137–178.

Chapter 4

Algorithms and Data Structures

4.1 Introduction

Algorithms are procedures for solving stated problems. Algorithm and computation have a long history, perhaps as long as civilization itself. The achievements of the ancient civilizations of China, Egypt, India, and Mesopotamia leave us in no doubt that their architects, astronomers, engineers, and merchants knew how to perform the computations associated with their professions. Much of the historical information was culled from an excellent Web site at the University of St. Andrews: www-history.mcs.st-and.ac.uk/history/.

The scholars of the Hellenic civilization systematized many of the achievements of their predecessors and added enormous contributions of their own. Athens, Alexandria, and other Hellenic cities contained "Academies" where scholars worked and taught together — forerunners of modern universities. Scholars, such as Euclid (about 325–265 BCE) and Eratosthenes (276–194 BCE), discovered or perfected a variety of algorithms that are still used today. Unfortunately, we have only fragmentary and indirect knowledge of the algorithms developed by these very ancient civilizations (Watt and Brown, 2001).

4.1.1 *Basic measurements for the performance of algorithms — Algorithm efficiency*

An algorithm is an automatic procedure for solving a stated problem, a procedure that could (at least in principle) be performed by a machine

(Watt and Brown, 2001). Concerning algorithms themselves and notation, we can state the following principles:

- The algorithm will be performed by some processor, which may be a machine or a human.
- The algorithm must be expressed in steps that the processor is capable of performing.
- The algorithm must eventually terminate, producing the required answer.
- The algorithm must be expressed in a language or notation that the processor "understands".

Often there exist several different algorithms that solve the same problem. We are naturally interested in comparing such alternative algorithms. Which is the fastest? Which needs the least memory? Fortunately, we can answer such questions in terms of the qualities of the algorithms themselves, without being distracted by the languages in which they happen to be expressed.

Given an algorithm, we are naturally interested in discovering how efficient it is. Efficiency has two distinct facets:

- **Time efficiency** is concerned with how much (processor) time the algorithm requires.
- **Space efficiency** is concerned with how much space (memory) the algorithm requires for storing data.

For telegeoprocessing system design and applications, we shall tend to pay more attention to time efficiency than to space efficiency. This is simply because time efficiency tends to be the critical factor in choosing between alternative algorithms.

Usually, the time taken by an algorithm depends on its input data. How should we measure an algorithm's time efficiency? Perhaps the most obvious answer is to use real time, measured in seconds. Real time is certainly important in many practical situations. Nevertheless, there are difficulties in using real time as a basis for comparing algorithms. An algorithm's real-time requirement depends on the processor speed as well as on the algorithm itself. Any algorithm can be made to run faster by using a faster processor, but this tells us nothing about the quality of the algorithm itself. And where the processor is a modern computer, the difficulty is compounded by the presence of software and hardware refinements — such as multiprogramming, pipelines, and caches — that increase the average speed of processing but make it harder to predict the time taken by an individual algorithm.

We prefer to measure an algorithm's time efficiency in terms of the algorithm itself. One idea is simply to count the number of steps taken by the algorithm until terminates. The trouble with this idea is that it depends on the granularity of the algorithm steps.

The most satisfactory way to measure an algorithm's time efficiency is to count characteristic operations. Which operations are characteristic depends on the problem to be solved.

If we want to understand the efficiency of an algorithm, we first choose characteristic operations and then analyze the algorithm to determine the number of characteristic operations performed by it. In general, the number of characteristic operations depends on the algorithm's input data.

For some algorithms, the analysis is straightforward. For other algorithms, the analysis is more complicated. We sometimes have to make simplifying assumptions, and we sometimes have to be content with estimating the maximum (or average) number of characteristic operations rather than the exact number.

Here is the heart of the matter. When we compare the efficiencies of alternative algorithms, what is most illuminating is the comparison of growth rates. Of course, we are interested in the actual times taken by alternative algorithms, but we are especially interested in the rates at which their time requirements grow with the number of input data, n. This interest in growth rates is easily justified. When n is small, we do not really care which algorithm is fastest because none of the algorithms will take much time. But when n is large, we certainly do care because all of the algorithms will take more time, and some might take much more time than others. *All else being equal, we prefer the algorithm whose time requirement grows most slowly with n.*

If we have a formula for an algorithm's time requirement, we can focus on its growth rate as follows:

- Take the fastest-growing term in the formula and discard all slower-growing terms.
- Discard any constant factor in the fastest-growing term.

The resulting formula is called the algorithm's time complexity. We define space complexity similarly.

We have now introduced the *O*-notation for algorithm complexity. The notation *O*(*X*) stands for "of order *X*" and means that the algorithm's time (or space) requirement grows proportionately to *X*. *X* characterizes the algorithm's growth rate, neglecting slower-growing terms and constant factors. In general, *X* depends on the algorithm's input data.

Table 4.1 summarizes the most common time complexities. The complexities are arranged in order of growth rate: *O*(1) is the slowest-growing and *O*(2*n*) is the fastest-growing. Figure 4.1 shows the comparison of

Table 4.1. The most common time complexities (Watt and Brown, 2001).

O(1)	Constant time	(Feasible)
O(log *n*)	Logarithmic time	(Feasible)
O(*n*)	Linear time	(Feasible)
O(*n* log *n*)	Log linear time	(Feasible)
O(*n*2)	Quadratic time	(Sometimes feasible)
O(*n*3)	Cubic time	(Sometimes feasible)
O(2*n*)	Exponential time	(Rarely feasible)

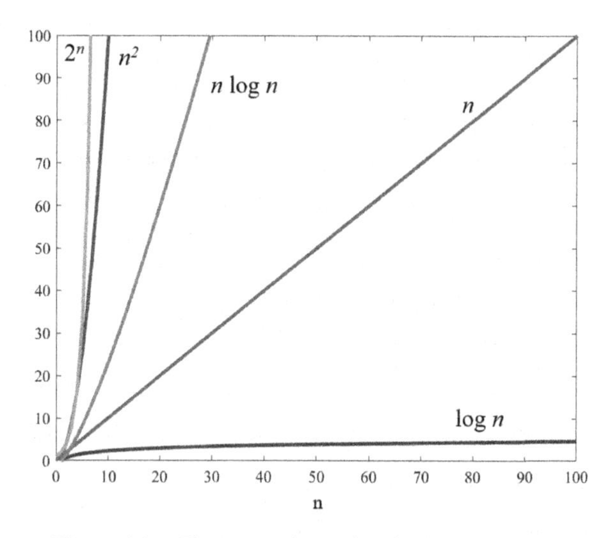

Figure 4.1. The comparison of various growth rates.

various growth rates graphically. We say that an algorithm is feasible if it is fast enough to be used in practice. Likewise, we say that a problem is feasible if it can be solved by a feasible algorithm.

More generally, $O(1)$ beats $O(\log n)$ beats $O(n)$ beats $O(n \log n)$ beats $O(n2)$ beats $O(n3)$ beats $O(2n)$.

4.1.2 *Basics of data structures and data types*

A data structure is a systemic way of organizing a collection of data. Arrays and linked lists are two simple and obliquitous data structures. More sophisticated data structures include binary search trees, hash tables, heaps, AVL trees, and B-trees. The most common Abstract Data Types (ADTs) are lists, sets, and maps. Other ADTs are stacks, queues, priority queues, and graphs.

A familiar example of a data structure is the array. The array is sometimes called a static data structure because its data capacity is fixed at the tie it is created. The dynamic data structures can freely expand and contract in accordance with the actual amount of data to be stored. Every data structure needs a variety of algorithms for processing the data contained in it.

A linked list consists of a sequence of nodes, connected by links. Each node contains a single element, together with links to one or both neighboring nodes.

Trees are also extremely important in computing. A binary tree is a group of nodes such that

- each node contains an element (value or object), together with links to at most two other nodes (its left child and its right child),
- the tree has a header, which contains a link to a designated node (the root node), and
- every node other than the root node is the left or right child of exactly one other node (its parent). The root node has no parent — the only link to it is the header.

Binary search trees (BSTs) are an effective representation of sets and maps. The BST search, insertion, and deletion algorithms have $O(log\ n)$ time complexity, provided that the BST remains well balanced. The weakness of BSTs is that all these algorithms degenerate to $O(n)$ time complexity if the BST becomes ill-balanced. An AVL tree is a BST that is height-balanced. AVL trees are named after the mathematicians G. M. Adel'son Vel'skii and Y. M. Landis, who invented them. The height of a node in a tree is the number of links that must be followed from that node to reach its remotest descendant. The height of a (sub-)tree is the height of its topmost node. A node is height-balanced if the heights of its sub-trees differ by at most one. (If it has only one sub-tree, the latter must have height 0, i.e., it must have a single node.) A tree is height-balanced if all its nodes are height-balanced.

The AVL tree search algorithm is identical to the BST search algorithm. The only difference — but a critical difference — is that the height-balanced property guarantees that the AVL tree search algorithm's time complexity is $O(log\ n)$.

A balanced search tree is an efficient data structure for representing a set or map for one fundamental reason: its depth increases slowly as the tree grows. Another way of saying the same thing is that the tree's size grows rapidly as the tree gets deeper.

A B-tree is a restricted form of k-ary tree, equipped with insertion and deletion algorithms that guarantee that the B-tree remains both shallow and well balanced as it grows and shrinks. A k-ary B-tree (or B-tree of order k) is a tree with the following properties:

- The root node contains at least 1 and at most k–1 elements. Every other node contains at least (k–1)/2 and at most k–1 elements.
- A non-leaf node that contains e elements ($1 \leq e < k$) also has links to exactly $e + 1$ child nodes. Each element sits between a pair of child links, which are links to its left child and right child, respectively.
- All leaf nodes lie at the same depth. A leaf node has no children.
- The elements in each node are arranged in ascending order. For each element elem in a non-leaf node, all elements in elem's left sub-tree are less than elem, and all elements in elem's right sub-tree are greater than elem.

The size of a B-tree is the total number of elements in that B-tree. The size of an individual node is the number of elements in that node. The empty B-tree has size zero. It has no nodes and no elements. The arity of a B-tree is the maximum number of children per node, i.e., k. In Java, we can represent a B-tree by an object of class Btree and each B-tree node by an object of class Btree.Node.

This data structure is simple and clear but rather wasteful space. Every node contains an array of child links, but in a leaf node, these child links are all null. An alternative would be to distinguish between leaf and non-leaf nodes and let only a non-leaf node contain an array of child links.

4.2 Spatial Data Structures

The growing importance of geographic user interfaces and off applications such as geo-information systems has confronted many application programmers with a challenging new task: processing large amounts of spatial data, off disk, correctly, and efficiently. As spatial data pose distinctly novel problems as compared to traditional "business data", the task is daunting in particular the following two (Nieverrrgelt and Widmayer, 1997):

- Access is primarily in proximity relations to other objects that populate Euclidean space rather than via inherent properties of objects, such as attribute values. A typical query is of the form "retrieve all objects that intersect a given region of space".
- Guaranteed correctness in processing spatial objects has proven to be a thorny problem requiring a systematic analysis of degenerate configurations. Moreover, efficiency depends on the interplay between two techniques: one or representing an object (independently of its location in space) and the other or storing the entire collection of objects in relation to their position in space.

The discipline of data structures, as a systematic body of knowledge, is truly a creation of computer science. Numerical computation in science and engineering mostly leads to linear algebra and hence matrix computations. Matrices are static datasets: the values change, but the shape and size of a matrix rarely do — this is true even for most sparse matrices, such as band matrices, where the propagation of non-zero elements is bounded. Array was Goldstine and von Newmann's answer to the requirement of random access.

The single most important issue that distinguishes spatial data structures from more traditional structures can be summarized in the phrase "organizing the embedding space versus organizing its contents". Data structures based on regular radix partitions of the space organize the domain from which data values are drawn in a systematic manner. Since they support the metric denied on this space, a prerequisite for efficient query processing, we call them metric data structures. The well-known quadtree (Samet, 1990a, 1990b) illustrates the advantages of a regular partition of the embedding space. A hierarchical partition of this square into quadrants and sub-quadrants, down to any desired level of granularity, provides a general-purpose scheme for organizing space, a skeleton to which any kind of spatial data can be attached for systematic access.

Spatial query processing is best described in terms of three steps (cell addressing, coarse filter, and fine filter) that can be analyzed independently (Nieverrrgelt and Widmayer, 1997). Data structures or external storage supports spatial queries to a set of geometric objects by realizing a fast but inaccurate filter: the data structure returns a set of external storage blocks that together contain the requested objects (hits) and others (false drops). Thereafter, a fine filter analyzes each object retrieved to either include or exclude it from the response. The purpose of a data structure is to associate each object with a disk block in such a way that the required operations are performed efficiently.

Two dominant factors guide the partition of the embedding space into cells, namely the data (the objects) and the space.

4.2.1 *Space driven partitions*

The most regular partitions of the data space are those that take into account only the amount of data to be stored (measured by the number of objects or by the number of storage blocks needed) but disregard the objects themselves and their specific properties. The number of objects merely determines the number of cells of the partition but not their location, size, or shape. The latter are inferred from a generic partition pattern for any number of regions desired. The typical examples are linear hashing, multidimensional linear hashing (Enbody and Hu, 1988), Dynamic z-hashing (Hutflesz *et al.*, 1988), space-filling curves (Bugnion *et al.*, 1997).

4.2.2 *Data driven partitions*

The most adaptive partitions of all are those defined by the set of data points. Since the partition tends to be less regular, a mechanism to keep track of the partition is needed. A natural choice for such a mechanism is a hierarchy, and hence multidimensional generalizations of one-dimensional tree structures have been proposed for that purpose. Prime examples are the k–d–B-tree (Robinson, 1981), a B-tree version (Comer, 1979) of the k–d-tree (Bentley, 1975), and a modified version of it, the hB-tree (Lomet and Salzberg, 1989, 1990).

4.2.3 *Combinations of space driven and data driven partitions*

In their simplest form, these partitions follow the generic pattern of space-driven partitions, with different levels of refinement across the data space, determined by the location of the geometric objects. A typical example of a one-dimensional data structure of this type is extendible hashing (Fagin *et al.*, 1979), where the data space is partitioned by recursively halving exactly those sub-spaces that contain too many data points. Other examples include EXCELL and the grid file (Tamminen, 1981, 1982), hierarchical grid files (Hinrichs, 1985; Krishnamurthy and Whang, 1985; Seeger and Kriegel, 1990), and the BANG file (Freeston, 1987).

4.2.4 *Spatial data structures for extended objects*

So far we have considered the simplest of spatial objects only, points, for the good reason that any spatial data structure must be able to handle point data efficiently. Most applications, however, deal with complex spatial objects. And although complex objects are composed of simpler building blocks, we encounter a multitude of the latter: line segments, poly-lines, triangles, aligned rectangles, simple polygons, circles and ovals, and the multidimensional generalizations of all these. Most data structures that support extended objects limit themselves to aligned rectangles (bounding boxes); exceptions include Bruzzone *et al.*, (1983); Günther, (1992); Jagadish, (1990). There are essentially three different extremal solutions to this problem, and the fourth one is a combination of two extremes. The three solutions are parameter space transformations (Henrich *et al.*, 1989; Hinrichs, 1985), clipping (Nguyen *et al.*, 1993; d'Amore *et al.*, 1995), and regions with unbounded overlap — R-tree (Guttman, 1984). Multilayer structure (Abel and Smith, 1983), R-file (Hutflesz *et al.*, 1990), and Guard file (Nievergelt and Widmayer, 1993) are the methods used in the fourth solution "cells with bounded overlap".

4.3 The Data Structure and Algorithms for GIS

GIS are complex systems consisting of a database part and a spatial data handling part. Spatial data handling in current GIS is less well-developed than the database aspects, and it is in relation to the spatial aspects that geometric algorithms and modeling can stimulate the process in GIS. GIS include many geometry-related problems such as spatial data storage and retrieval, visualization of spatial data, overlay of maps, spatial interpolation, and generation. Van Kreveld *et al.* (1997) summarized the two lines of research whose interaction promises to have significant practical impact in the near future: the application-oriented discipline of GIS and the technical discipline of geometric computation, or in particular, geometric algorithms, and spatial data structures.

Algorithm research has been an integral part of computer science since its beginnings. The goal is to design well-defined procedures that solve well-defined problems and analyze their efficiency. The motivation for studying geometric algorithms comes from areas like robotics, computer graphics, automated manufacturing, VLSI design, and GIS.

The collaboration of GIS research and computational geometry has intensified in recent years, for several reasons. First, the availability of geographic data in digital form is increasing rapidly. The acquisition of geographic data and the digitizing of maps are laborious, time-consuming tasks, and without appropriate digital data, GIS cannot operate effectively. Second, improved hardware now makes it possible to store and process large amounts of geographic data and to produce high-quality maps on computers. Third, the research community in computational geometry, after having laid the conceptual and technical foundation, has recently shifted its attention toward more practical and more applied solutions.

Many standard geometric problems and solutions, even those without any explicit reference to geography, are useful to GIS. Storing polygonal sub-divisions on primary or background storage for efficient windowing queries is one example. Implementation and testing of data structures and query algorithms reveal their usefulness to GIS. Others include topological data structures like the doubly connected edge list or quad edge structure, spatial data structures, like the quadtree and R-tree, the Voronoi diagram, the Delaunay triangulation, map overlay and buffer computation, and visualization algorithms. They are well documented in textbooks or other standard texts.

Most of the basic algorithms developed in computational geometry, however, are not directly applicable to GIS. This is because standard geometric problems are often simplified to such an extent that they neglect important requirements of GIS. In addition, many problems arising in GIS have not yet been formalized in sufficient detail for the design of efficient algorithms.

4.3.1 *Voronoi methods in GIS*

Access to spatial data is primarily guided by proximity relations among objects. Traditional vector-based GIS organism basic objects of interest such as roads, rivers, towns, or houses in thematic multilayered (polygonal) maps representing them as polygons, arcs, and nodes. Much of the early work was based on the line-intersection model of space where the global detection of intersecting arcs or lines was the sole method of determining connectivity — see Dutton (1978). Relations between polygons, arcs, and nodes imposed a "topology" which linked the basic map objects. All operations for queries of polygonal thematic maps in GIS systems are global and depend therefore on a complete rebuild of the topology after each operation.

Many weaknesses of this technique have been identified. The detection of intersections itself is both an expensive operation and prone to errors due to digitizing limitations; with the consequence that if an adjacency is defined via intersection, disconnected features such as islands are difficult to handle correctly. As a result, there is no readily available approach to locally modify the topology and hence no true dynamic system permitting the addition and deletion of individual map objects.

One potential solution is readily available from computational geometry: the Voronoi diagram, a universal data structure for representing proximity (Gold *et al.*, 1997). It supports a multitude of nearest neighbor queries (Okabe *et al.*, 1992; Preparata and Shamos, 1985) as we will see, the Voronoi diagram for points and line segments with its dual, the Delaunay multigraph (Delaunay, 1934), provides the basic spatial adjacency properties between map objects (Figure 4.2). The Voronoi diagram allows us to locate the nearest generator to a given query point whereas the dual gives us fast access to the neighbors of a generator stored in the GIS. This not only resolves the basic difficulties with the line-intersection model but allows us to maintain the topology of a GIS.

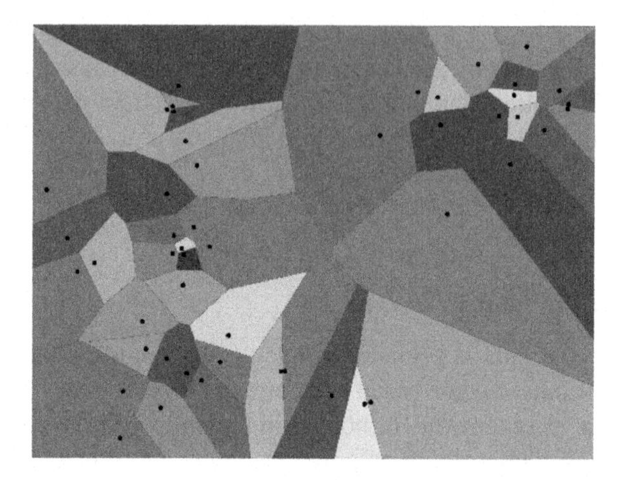

Figure 4.2. Voronoi graph.

Nowadays, the Voronoi diagram and its variants are well known in many areas of science (Aurenhammer, 1991) and many algorithmic paradigms of computational geometry apply to compute them efficiently (Boissonnat, 1992; Fortune, 1987; Guibas *et al.*, 1990; Shamos and Hoey, 1975). In geomatics and GIS, Gold (1990, 1991, 1994) discovered the Voronoi diagram as a fundamental tool for representing and maintaining topology in a map. Gold *et al.* (1997) gave a survey of static, kinematic, and dynamic Voronoi diagrams as basic tools for GIS. They presented a method that allows the insertion, deletion, and translation of points and line segments in a Voronoi diagram of n generator. All elementary operations are available in $O(log\ n)$ expected time and linear expected storage space. The Voronoi approach also greatly simplifies some of the basic traditional GIS queries and allows even new types of higher-level queries.

Within the Voronoi diagram, a map polygon creates Voronoi cells in its interior. There are three main data types stored in a Voronoi system: coordinates, Delaunay triangles (containing pointers to the three adjacent triangles and to the three map objects forming the vertices), and the map objects themselves, which may be points or open line segments (and have pointers, as required, to the coordinate records and the matching line segment). No pointers are associated with the map objects themselves; instead, all topology is contained within the Delaunay triangle records. This is an extremely desirable property in a GIS, as the same object may then be inserted in several layers simultaneously. This eliminates object duplication in the attribute database, as well as avoids having different coordinate representations of the same object in various layers.

For real-world digitizing, corrections are formed when the line being drawn crosses a previously defined line. In this case, intersections must be detected in advance of the moving point. Gold (1992) showed that a line segment can only be intersected by an expanding line segment if both have been Delaunay neighbors shortly before. So, if a line segment becomes a new Delaunay neighbor of an expanding line segment, an intersection test is made and, if necessary, the intersection point is calculated, and both the intersected line segment and the expanding line segment are split off at the intersection point as described above. This gives a fully dynamic algorithm for the creation and deletion of points and line segments, for point movement, and for intersection and the snapping together of lines during digitizing.

When maps become very large — maybe even global — additional issues become critical, e.g., partitioning the map into portions that are manageable in memory becomes a significant issue. Traditional GIS, with map sheet boundaries, do not have a good answer. Some suggestions for partitioning maps along line segment boundaries or constrained triangulation edges are given in Yang and Gold (1996). This has the added attraction that such a partitioning may be used to control access to particular portions of the map, either so that several operators may simultaneously be working on map updates within assigned regions or else to allow parallel processing to build the whole map more quickly without danger of conflict. Splitting and merging Voronoi diagram as described above are fundamental in this process (Gold *et al.*, 1997).

4.3.2 *Digital elevation models*

Two of the most important types of maps are the choropleth map and the isoline map (van Kreveld, 1997). A choropleth map is basically a subdivision into regions, where the boundaries separate regions with a different attribute or property. For example, on a map showing countries the boundaries always have different countries on the two sides. An isoline map is also a sub-division into regions, but now the boundaries show all locations where an attribute has a fixed value. A precipitation map showing the lines of 750 mm, 800 mm, 850 mm, and 900 mm of precipitation is an example.

Isoline maps are a way to visualize elevation data, which is represented in a GIS as a digital elevation model (DEM). Mathematically, an elevation model is a continuous function in two variables. A digital elevation model simply is a finite representation of an elevation model. The most well-known example of an elevation mode is height above sea level.

In the GIS literature, papers abound on the application and representation of digital elevation models. Review papers have been written as well (Weibel and Heller, 1991), and textbooks on GIS and automated cartography also deal with digital elevation models (Clarke, 1995; Kraak and Ormeling, 1996; Worboys, 1995; Maguire *et al.*, 1991). The most notable survey paper with emphasis on algorithms on DEM was from van Kreveld (1997). It is almost impossible to produce a complete survey on all DEM algorithms or even a nearly complete bibliography. In this book, we will highlight several most important concepts and problems on DEM data and discuss a few algorithms more thoroughly. The emphasis is on the efficiency of the algorithms, but it appears that "model" and "algorithm" cannot be seen as separate issues. The algorithms that perform the computation cannot really be compared on efficiency: the algorithms don't compute the same thing (van Krevveld, 1997). Even worse, the algorithms may be based on different terrain models. Another issue of importance is how the efficiency of algorithms should be analyzed. Computational geometers usually consider the worst possible inputs and make sure that the algorithm works well even in these cases. GIS researchers often don't analyze their algorithms or give timings of an implementation.

4.3.2.1 *The regular square grid terrain models and representation*

The regular square grid — or simply the grid — is a structure that species values at a regular square tessellation of the domain. In a computer, it is stored as a two-dimensional array. For every square in the tessellation, or entry in the array, exactly one elevation value is specified.

4.3.2.2 *The contour line model*

In the contour line model to represent elevation, some set of elevation values is given, and every contour line with one of these elevations is represented in the model. So, a collection of contour lines is stored along with its elevation. Sometimes, the term isoline or isocontour is used when the elevation model represents something else than height above sea level. A contour line is usually stored as a sequence of points with its x and y coordinates. It then represents a simple polygon or polygonal chain of which the elevation is specified. The contour lines can be stored in a doubly connected edge list (de Berg *et al.*, 1997) or quad edge structure (Guibas and Stolfi, 1985). An alternative to this representation is by means of the contour tree (Freeman and Morse, 1967) or topographic change tree (Kweon and Kanade, 1994) or Reeb graph (Takahashi *et al.*, 1995).

4.3.2.3 *The triangulated irregular network (TIN) model*

In the triangulated irregular network (TIN) model, usually abbreviated as TIN, a finite set of points is stored together with their elevation. The points need not lie in any particular pattern, and the density may vary. On these points, a planar triangulation is given. Any point in the domain will lie on a vertex, an edge, or in a triangle of the triangulation. A possible storage scheme of a TIN is the doubly connected edge list (Preparata and Muller, 1979; Preparata and Shamos, 1985) or the quad edge structure (Guibas and Stolfi, 1985). Both are ways of representing the topology of any planar sub-division. These representations allow for all necessary traversal operations in an efficient manner.

4.3.2.4 *Hierarchical models*

A hierarchical terrain model is a terrain model that represents a terrain in various levels of error or, to make it sound less bad, various levels of impression. Most approaches of this type are based on TINs. A hierarchical terrain model allows the user to choose a terrain with the appropriate precision for each task. A survey on the topic exists (De Floriani *et al.*, 1994).

4.3.3 *Access to a TIN*

In a regular square grid structure, there is direct access to every part of the terrain. If one wants to know the elevation at a point with coordinates x

and *y*, these coordinates can simply be rewritten to index values in the two-dimensional array. In a TIN, this is not so easy because a TIN is a pointer structure. If one wants to know the elevation at a point given its coordinates, one could test each triangle to see if it contains the point. This is rather inefficient, obviously. There are three methods to gain access to the TIN at a specific point: access using quadtrees (Samet, 1990a), access using planar point location (Sarnak and Tarjan, 1986; Seidel, 1991), and jump-and-walk strategy (Mücke *et al.*, 1996). The simplest method to locate a point in a TIN among these three is the jump-and-walk strategy. It also requires very little additional storage, so it may well be the best choice in practice.

4.3.4 *Conversion between terrain model*

Terrain data can be entered into a GIS in various formats. Often, contour line data are entered when paper maps with contour lines are digitized by hand. Also quite often, gridded data are the input format, for instance, when the data are acquired by remote sensing or automatic photo-interpretation. Gridded data usually are huge in size, resulting in high memory requirements and slow algorithms when the data are processed further. Contour line data often need to be interpreted anyway before anything useful can be done with it.

Suppose a set *P* of *n* points in the plane is given, each with an elevation value. To convert this information into a TIN, one could simply triangulate the point set. It is common to use the Delaunay triangulation because it attempts to create well-shaped triangles (de Berg *et al.*, 1997). When the interpolation provided by the Delaunay triangulation is not appropriate, one can use different, more advanced interpolation methods like natural neighbor interpolation (Sambridge *et al.*, 1995), weighted moving averages (Burrough, 1986), splines (Foley *et al.*, 1990), or Kriging (Webster and Oliver, 1990).

Grid-to-TIN conversion can be seen as a special case of the conversion of sample points to a TIN. Also, grid-to-TIN conversion can be seen as a special case of TIN generation: reducing the number of vertices of a TIN to represent a terrain. A grid can simply be triangulated to a fine regular triangulation (Garland and Heckbert, 1995; Heller, 1990; Lee, 1989, 1991).

Contour line to TIN conversion algorithms is useful because elevation data are often obtained by digitizing contour line maps. To convert the contour lines to a TIN, the obvious thing to do is triangulate all regions, that is, triangulate between the contour lines. Instead of using any triangulation, it is a good idea to use one that gives nicely shaped triangles, like the Delaunay triangulation. It is called the constrained Delaunay triangulation (Chew, 1989; De Floriani and Puppo, 1989).

4.3.5 *Visualization*

One of the fundamental tasks any GIS has to perform is the visualization of geographic data. Often the data will be elevation data describing a terrain. There are two important topics that arise in the visualization of TINs (de Berg, 1997). The first is the hidden-surface-removal problem: given a TIN and a viewpoint (or a viewing direction), we want to compute what we see of the TIN when we look at it from the given viewpoint (or in the given viewing direction). The second topic concerns the following. The number of triangles in a TIN is often so large that it becomes infeasible to visualize all triangles in a reasonable amount of time. Fortunately, this is not necessary, as large parts of the TIN will be far from the viewpoint; these parts can be visualized at a lower level of detail, using less triangles. To make this work, one should pre-compute representations of the same TIN at various levels of detail (LOD) (Sun *et al.*, 2002).

There are two major approaches to perform hidden-surface removal (Sutherland *et al.*, 1974). On eyes to first determine which part of each object is visible and then project and scan-convert only the visible parts. Algorithms using this approach are called object-space algorithms. Dorward (1994) gave an extensive overview of output-sensitive hidden-surface-removal algorithms, and de Berg's book (1993) also contains an ample discussion of object-space hidden-surface removal.

The other possibility is to first project the objects and decide during the scan-conversion for each individual pixel which object is visible at that pixel. Algorithms following this approach are called image-space algorithms. The Z-buffer algorithm, which is the algorithm most commonly used to perform hidden-surface removal, uses this approach. The Z-buffer algorithm is easy to implement, and any graphics workstation provides it, often in hardware (Booth *et al.*, 1986).

Image-space methods compute the view of a scene pixel by pixel. This means that the “structure” of the view is lost. Object-space algorithms compute a combinatorial representation of the view. In general, object-space hidden-surface-removal methods tend to be slower than image-space methods such as the Z-buffer algorithm because they cannot be implemented in hardware very well (de Berg, 1997).

For a realistic representation of a terrain, millions of triangles are needed. In applications such as flight simulation, a terrain should be rendered at real time, but even with modern technology, it is impossible to achieve this when the number of triangles is this large. Fortunately, a realistic image of the terrain is only crucial when one is close to the terrain and only a small part of the terrain is visible; when one is flying high above the terrain, a coarse representation suffices. So, what is needed is a multiresolution model: a hierarchy of representations at various levels of detail. This makes it possible for a given viewpoint to render the terrain at an adequate level of detail. The idea of multiresolution models is quite old, and work on it is scattered over literature in graphics, GIS, and other areas. Heckbert and Garland (1994) and De Floriani *et al.* (1994) gave nice surveys of many of the existing multiresolution techniques.

The best-known hierarchical data structure is probably the quadtree (Samet, 1990a, 1990b). Although quadtrees are based on regular grids, not on TINs, it has been widely used as a multiresolution model (de Berg, 1997).

In computational geometry, triangulation is performed for building efficient data structures based on the decomposition of a polygon into regular pieces. Such data structures are used for many applications including processing shortest path queries and ray shooting. On the other hand, in almost all applications in computer graphics and geographic information systems (GIS), the main requirement is the fast rendering of a polygon. A trapezoid is as simple a shape as a triangle and it is easy to render like a triangle in raster graphics devices. In many GIS applications, it is often necessary to render many GIS polygons with a large number of vertices simultaneously. Common graphics APIs are usually slow in rendering such large polygons.

The decomposition of a simple polygon into triangles is a well-studied field of computer graphics (Bern, 1997) and computational geometry. In computer graphics and GIS applications, it is comparatively easier to render a simple shape like a triangle than an arbitrary polygon. Similarly, in computational geometry, a large number of data structures related to polygons are based on triangulation of those polygons. For many triangulation algorithms, a *trapezoidal decomposition* of the polygon is computed as the first step to *triangulation*. A trapezoid is a four-sided polygon with at least two parallel sides. The key difference between a trapezoidal decomposition and a triangulation is that the vertices defining a particular trapezoid can be additional points on the boundary of the polygon, whereas, in triangulation, it is not allowed to introduce new vertices.

One of the first triangulation algorithms was designed by Garey *et al.* (1978). Their algorithm is based on the method of dividing the input polygon into monotone pieces and it runs in $O(n \log n)$ time. The best-known triangulation algorithm is by Chazelle (1991) and it runs in $O(n)$ time, where n is the number of vertices of the polygon. However, Chazelle's algorithm is extremely complicated and is not practical for implementation. Even now, the algorithm by Garey *et al.* (1978) is extensively used for its simplicity and programming ease. Chazelle and Incerpi (1984) improved upon this algorithm by introducing the notion of *sinuosity* which is the number of *spiraling* and *anti-spiraling* polygonal chains on the polygon boundary. Their algorithm runs in $O(n \log s)$ time, where s is the sinuosity of the polygon. The sinuosity of a polygon is usually a small constant in most practical cases and the algorithm (Chazelle and Incerpi, 1984) is an almost linear-time algorithm. However, the algorithm by Chazelle and Incerpi (1984) is based on a divide-and-conquer strategy and is still quite complex to implement.

The fastest known algorithm for trapezoidation of general GIS datasets is by Lorenzetto *et al.* (2002). Their algorithm improves upon a previously published algorithm by Žalik and Clapworthy (1999) and can perform a trapezoidal decomposition of general GIS polygons with holes in $O(n \log n)$ time, where n is the total number of vertices defining the polygon and all holes inside it. However, many GIS polygons do not contain holes inside them and it is interesting to investigate whether it is possible to design a simpler and faster algorithm for decomposing such polygons.

References

Abel, D. J. and Smith, J. L. (1983). A data structure and algorithm based on a linear key for a rectangle retrieval problem. *Computer Vision, Graphics, and Image Processing*, **24**: 1–13.

Aurenhammer, F. (1991). Voronoi diagrams: A survey of a fundamental geometric data structure. *ACM Computing Survey*, **23**: 345–405.

Bentley, J. L. (1975). Multidimensional binary search used for associative searching. *Communications of the ACM*, **18**(9): 509–517.

Bern, M. (1997). *Handbook of Discrete and Computational Geometry*, Chapter 22, pp. 413–428. Boca Raton: CRC Press.

Boissonnat, J. D., Devillers, O., Schott, R., Teillaud, M., and Yvinec, M. (1992). Applications of random sampling to on-line algorithms in computational geometry. *Discrete & Computational Geometry*, **8**: 51–71.

Booth, K. S., Forsey, D. R., and Paeth, A. W. (1986, November). Hardware assistance for Z-buffer visible surface algorithms. *IEEE Computer Graphics and Applications*, **6**(11): 31–39.

Bruzzone, E., De Floriani, L., and Pellegrini, M. (1983). A hierarchical spatial index for cell complexes. In *Proceedings of 3rd International Symposium on Advances on Spatial Databases*, Singapore. Lecture Notes in Computer Science, Vol. 692, pp.105–122. Berlin: Springer-Verlag.

Bugnion, E., Roos, T., Wattenhofer, R., and Widmayer, P. (1997). Space filling curves versus random walks. In *Lecture Notes in Computer Science*, Vol. 1340, pp. 199–212. Berlin: Springer-Verlag.

Burrough, P. A. (1986). *Principles of Geographical Information Systems for Land Resources Assessment*. New York: Oxford University Press.

Chazelle, B. (1991). Triangulating a simple polygon in linear time. *Discrete and Computational Geometry*, 6: 485–524.

Chazelle, B. and Incerpi, J. (1984). Triangulation and shape complexity. *ACM Transactions on Graphics*, **3**: 153–174.

Chew, L. P. (1989). Constrained delaunay triangulations. *Algorithmica*, **4**: 97–108.

Clarke, K. C. (1995). *Analytical and Computer Cartography*, 2nd ed. Englewood Cliffs, NJ: Prentice Hall.

Comer, D. (1979). The ubiquitous B-tree. *ACM Computing Surveys*, **11**(2): 121–138.

D'Amore, F., Nguyen, V. H., Roos, T., and Widmayer, P. (1995). On optimal cuts of hyperrectangles. *Computing*, **55**: 191–206. Springer-Verlag.

de Berg, M. (1993). *Ray Shooting, Depth Orders and Hidden Surface Removal*. Lecture Notes in Computer Science, Vol. 703. Berlin: Springer-Verlag.

de Berg, M. (1997). Visualization of TINs. In van Kreveld, M., Nievergelt, J., Roos, T., and Widmayer, P. (Eds.), *Algorithmic Foundations of Geographic Information Systems*. Lecture Notes in Computer Science, Vol. 1340, pp. 79–98. Berlin: Springer-Verlag.

de Berg, M., van Kreveld, M., Overmars, M., and Schwarzkopf, O. (1997). *Computational Geometry — Algorithms and Applications*. Berlin: Springer-Verlag.

De Floriani, L. and Puppo, E. (1989). A survey of constrained Delanuary triangulation algorithms for surface representation. In Pieroni, G. G. (Ed.), *Issues on Machine Vision*, pp. 95–104. New York: Springer-Verlag.

De Floriani, L., Marzano, P., and Puppo, E. (1994). Hierarchical terrain models: survey and formalization. In *Proceedings of ACM Symposium on Applied Computing*, 1994.

Delaunay, B. N. (1934). Sur la sphere vide, *Bull. Acad. Science USSR VII: Class. Science Math.*, pp. 793–800.

Dorward, S. E. (1994). A survey of object-space hidden surface removal. *International Journal of Computer Geometry Applications*, **4**: 325–362.

Dutton, G. (Ed.). (1978). *First International Advanced Symposium on Topological Data Structures for GIS*, 1978, Vol. 8. Cambridge, MA: Harvard University.

Enbody, R. J. and Du, H. C. (1988). Dynamic hashing schemes. *ACM Computing Surveys*, **20**(2): 85–113.

Fagin, R., Nievergelt, J., Pippenger, N., and Strong, H. R. (1979). Extendible hashing — A fast access method for dynamic files. *ACM Transactions on Database Systems*, **4**(3): 315–344.

Foley, J. D., van Dam, A., Feiner, S. K., and Hughes, J. F. (1990). *Computer Graphics: Principles and Practice*. Reading, MA: Addison-Wesley.

Fortune, S. (1987). A sweeping algorithm for Voronoi diagrams. *Algorithmica*, **2**: 153–174.

Freeman, H. and Morse, S. P. (1967). On searching a contour map for a given terrain elevation profile. *Journal of the Franklin Institute*, **284**: 1–25.

Freeston, M. W. (1987). The BANG file: A new kind of grid file. In *Proceedings of ACM SIGMOD International Conference on the Management of Data, San Francisco*, pp. 260–269.

Garey, M., Johnson, D., Preparata, F., and Tarjan, R. (1978). Triangulating a simple polygon. *Information Processing Letters*, **7**: 175–180.

Garland, M. and Heckbert, P. S. (1995). Fast polygonal approximation of terrains and height fields. Technical Report CMU-CS-95-181, Carnegie Mellon University, 1995.

Gold, C. M. (1990). Spatial data structures — The extension from one to two dimensions. In L. F. Pau (Ed.), *Mapping and Spatial Modelling for Navigation*. NATO ASI Series F, No. 65, pp. 11–39. Berlin: Springer-Verlag.

Gold, C. M. (1991). Problems with handling spatial data — The Voronoi approach. *CISM Journal*, **45**(1): 65–80.

Gold, C. M. (1994). The interactive map. In Molenaar, M. and de Hoop, S. (Eds.), *Advanced Geographic Data Modelling and Query Languages for 2D and 3D Applications*, Netherlands Geodetic Commission, Publications on Geodesy, No. 40, pp. 121–128.

Gold, C. M., Remmele, P. R., and Roos, T. (1997). *Voronoi methods in GIS. Lecture Notes in Computer Science,* Vol. 1340, pp. 21–35. Berlin: Springer-Verlag.

Guibas, L. J. and Stolfi, J. (1985). Primitives for the manipulation of general subdivisions and the computation of Voronoi diagrams. *ACM Transactions on Graphics*, **4**: 74–123.

Guibas, L. J., Knuth, D. E., and Sharir, M. (1990). Randomized incremental construction of Delaunay and Voronoi diagrams. In *Proceedings of 17th International Colloquium on Automata, Languages and Programming ICALP'90*. LNCS. Vol. 443, pp. 414–431. Berlin: Springer-Verlag.

Günther, O. (1992). Evaluation of spatial access methods with oversize shelves. In *Geographic Data Base Management Systems, ESPRIT Basic Research Series Proceedings*, pp. 177–193. Berlin: Springer-Verlag.

Guttman, A. (1984). R-tree: A dynamic index structure for spatial searching. In *Proceedings of ACM SIGMOD International Conference on the Management of Data*, Boston, pp. 47–57.

Heckbert, P. S. and Garland, M. (1994). Multi-resolution modelling for fast rendering. In *Proceedings of Graphic Interface'94*, pp. 43–50. Information Processing Society.

Heller, M. (1990). Triangulation algorithms for adaptive terrain modelling. In *Proceedings of 4th International Symposium on Spatial Handling*, pp. 163–174.

Henrich, A., Sic, H. W., and Widmayer, P. (1989). The LSD-tree: Spatial access to multidimensional point- and non-point objects. In *Proceedings 15th International Conference on Very Large Data Bases*, Amsterdam, pp. 45–53.

Hinrichs, K. H. (1985). The grid file system: Implementation and case studies of applications. Dissertation, ETH Zurich.

Hutflesz, A., Six, H. W., and Widmayer, P. (1988). Globally order preserving multidimensional linear hashing. In *Proceedings of 4th International Conference on Data Engineering*, Los Angeles, pp. 572–579.

Hutflesz, A., Six, H. W., and Widmayer, P. (1990). The R-file: An efficient access structure for proximity queries. In *Proceedings of 6th International Conference on Data Engineering*, Los Angeles, pp. 372–379.

Jagadish, H.V. (1990). Spatial search with polyhedra. In *Proceedings of 6th International Conference on Data Engineering*, Los Angeles, pp. 311–319.

Kraak, M. J. and Ormeling, F. J. (1996). *Cartography — Visualization of Spatial Data*. Longman: Harlow.

Krishnamurthy, R. and Whang, K. Y. (1985). Multilevel grid files. IBM T. J. Watson Research Center Report, Yorktown Heights, New York.

Kweon, I. S. and Kanade, T. (1994). Extracting topographic terrain features from elevation maps. *CVGIP: Image Understanding*, **59**: 171–182.

Lee, J. (1989). A drop heuristic conversion method for extracting irregular network for digital elevation models. In *Proceedings of GIS/LIS'89*, November 1989, Vol. 1, pp. 30–39. American Congress on Surveying and Mapping.

Lee, J. (1991). Comparison of existing methods for building triangular irregular network models of terrain from grid digital elevation models. *International Journal of Geographic Information System*, **5**(3): 267–285.

Lomet, D. B. and Salzberg, B. (1989). A robust multi-attribute search structure. In *Proceedings of 5th International Conference on Data Management*, Los Angeles, pp. 296–304.

Lomet, D. B. and Salzberg, B. (1990). The hB-tree: A multiattribute indexing method with good guaranteed performance. *ACM Transactions on Database Systems*, **15**(4): 625–658.

Lorenzetto, G., Datta, A. and Thomas, R. (2002). A fast trapezoidation technique for planar polygons. *Computers & Graphics*, **26**: 281–289.

Maguire, D. J., Rhind, D. W., and Goodchild, M. F. (1991, November). *Geographical Information Systems*. Longman Science & Technology.

Mücke, E. P., Saias, I., and Zhu, B. H. (1996). Fast randomized point location without preprocessing in two- and three-dimensional delaunay triangulations. In *Proceedings of 12th Annual ACM Symposium on Computational Geometry*, pp. 274–283.

Nguyen, V. H., Roos, T., and Widmayer, P. (1993). Balanced cuts of a set of hyperrectangles. In *Proceedings of ACM SIGMOD International Conference on the Management of Data*, San Francisco, pp. 270–277.

Nievergelt, J. P. and Widmayer, P. (1993). Guard files: Stabbing and intersection queries on fat spatial objects. *The Computer Journal*, **36**(2): 107–116.

Nievergelt, J. P. and Widmayer, P. (1997). Spatial data structure: Concepts and design choices. In van Kreveld, M., Nievergelt, J., Roos, T. and Widmayer, P. (Eds.), *Algorithmic Foundations of Geographic Information Systems*, pp. 153–198. Berlin: Springer-Verlag.

Okabbe, A., Boots, B., and Sugihara, K. (1992). *Spatial Tessellations — Concepts and Applications of Voronoi Diagrams*. Chichester: John Wiley and Sons.

Preparata, F. P. and Shamos, M. I. (1985). *Computational Geometry: An Introduction*. New York: Springer-Verlag.

Preparata, F. P. and Muller, D. E. (1979). Finding the intersection of n half-spaces in time *O(n log n)*. *Theoretical Computer Science,* **8**: 45–55.

Robinson, J. T. (1981). The K-D-B-tree: A search structure for large multidimensional dynamic indexes. In *Proceedings of ACM SIGMOD International Conference on the Management of Data*, Ann Arbor, pp. 10–18.

Sambridge, M., Braun, J., and McQueen, H. (1995). Geophysical parameterization and interpolation of irregular data using natural neighbours. *Geophysical Journal International*, **122**: 837–857.

Samet, H. (1990a). *The Design and Analysis of Spatial Data Structure*. Reading: Addison-Wesley.

Samet, H. (1990b). *Applications of Spatial Data Structure*. Reading: Addison-Wesley.

Sarnak, N. and Tarjan, R. E. (1986). Planar point location using persistent search trees. *Communications of the ACM*, **29**: 669–679.

Seeger, B. and Kriegel, H. P. (1990). The buddy-tree: An efficient and robust access method for spatial data base systems. In *Proceedings of 16th International Conference on Very Large Data Bases*, Brisbane, pp. 590–601.

Seidel, R. (1991). A simple and fast incremental randomized algorithm for computing trapezoidal decompositions and for triangulating polygons. *Computational Geometry: Theory and Applications*, **1**: 51–64.

Shamos, M. I. and Hoey, D. (1975). Closest point problems. In *Proceedings of 16th Annual IEEE Symposium on Foundations of Computers Science FOCS'75*, pp. 151–162.

Sun, M., Xue, Y. and Ma, A.-N. (2002). *3D Visualization of Large DEM Data Set Based on Grid Division*. Lecture Notes in Computer Science, Vol. 2331, pp. 975–983.

Sutherland, I. E., Sproull, R. F., and Schumacker, R. A. (1974). A characterization of ten hidden-surface algorithms. *ACM Computer Survey*, **6**(1): 1–55.

Takahashi, S., Ikeda, T., Shinagawa, Y., Kunii, T. L., and Ueda, M. (1995). Algorithms or extracting correct critical points and constructing topological graphs from discrete geographical elevation data. In *Eurographics'95*, Vol. 14, pp. C-181–C-192.

Tamminen, M. (1981). The EXCELL method for efficient geometric access to data. Acta Polytechnica Scandinavica, Mathematics and Computer Science Series No. 34, Helsinki, Finland.

Tamminen, M. (1982). The extendible cell method for closest point problems. *BIT*, **22**: 27–41.

van Kreveld, M., Nievergelt, J., Roos, T., and Widmayer, P. (Eds.). (1997). *Algorithmic Foundations of Geographic Information Systems*. Lecture Notes in Computer Science, Vol. 1340. Berlin: Springer-Verlag.

van Kreveld, M. (1997). Digital elevation models and TIN algorithms. In van Kreveld, M., Nievergelt, J., Roos, T., and Widmayer, P. (Eds.), *Algorithmic Foundations of Geographic Information Systems*. Lecture Notes in Computer Science, Vol. 1340, pp. 37–78. Berlin: Springer-Verlag.

Watt, D. and Brown, D. (2001). *Java Collections: An Introduction to Abstract Data Types, Data Structures and Algorithms*. Hoboken: John Wiley.

Webster, R. and Oliver, M. A. (1990). *Statistical Methods in Soil and Land Resource Survey*. New York: Oxford University Press.

Weibel, R. and Heller, M. (1991). Digital terrain modelling. In Maguire, D. J., Goodchild, M. F. and Rhind, D. W. (Eds.), *Geographic Information Systems — Principles and Applications*, pp. 269–297. London: Longman.

Worboys, M. F. (1995). *GIS: A computing Perspective*. London: Taylor & Francis.

Yang, W. and Gold, C. M. (1996). Managing spatial objects with the VMO-Tree. In *Proceedings of 7th International Symposium on Spatial Data Handling SDH'96*, Delft, The Netherlands, August, 1996, Vol. 2, pp. 11B-15–11B-30.

Žalik, B. and Clapworthy, G. J. (1999). A universal trapezoidation algorithm for planar polygons. *Computers & Graphics*, **23**: 353–363.

Chapter 5

Grid Computing and Geoinformation

5.1 Introduction

What is the GRID? The idea behind it is simple. As very few people or organizations use the full capacity of their computers' processing power, the idea is to harness this unused resource through the internet to solve major computational problems that require far more "memory" and computing power than that available at any one site, rather in the same way that we plug into the electricity grid to run an electrical appliance whenever needed.

A scaled-down version of this idea was first put into practice a year ago by the Search for Extra Terrestrial Intelligence (SETI) project (http://setiathome.ssl.berkeley.edu/). By enlisting the help of volunteers, they are able to "plug into" the processing power of millions of home computers to sift through radio signals to search for messages from space.

The first modern Grid is generally considered to be the I-WAY, developed as an experimental demonstration project for SC95 (DeFanti *et al.*, 1996). In 1995, during the week-long Supercomputing conference, pioneering researchers came together to aggregate a nationally distributed testbed with over 17 sites networked together by the vBNS. Over 60 applications were developed for the conference and deployed on the I-WAY, as well as a rudimentary Grid software infrastructure (Foster *et al.*, 1997) to provide access, enforce security, coordinate resources, and other activities. Developing infrastructure and applications for the I-WAY provided a seminal and powerful experience for the first generation of modern Grid researchers and projects. This was important as the development of Grid research involves a very different focus than distributed computing research. Whereas distributed computing research generally focuses on addressing the problems of geographical separation, Grid research focuses on addressing the problems of integration and management of software.

I-WAY opened the door for considerable activity in the development of Grid software. The Globus (Foster and Kesselman, 1999) and Legion (Lewis and Grimshaw, 1995; Grimshaw and Wulf, 1996) infrastructure projects explored approaches for providing basic system-level Grid infrastructure. The Condor project (Frey *et al.*, 2001; Litzkow, 1998) experimented with high-throughput scheduling while the AppLeS (Casanova, 2000), Mars (Gehring and Reinefeld, 1996), and Prophet (Weissman, 1999) projects experimented with high-performance scheduling. The Network Weather Service (http://nws.cs.ucsb.edu/) project focused on resource monitoring and prediction, while the Storage Resource Broker (http://www.npaci.edu/DICE/SRB/) focused on uniform access to heterogeneous data resources. The NetSolve (http://icl.cs.utk.edu/netsolve/) and Ninf (http://ninf.apgrid.org/) projects focused on remote computation via a client-server model. These and many other projects provided a foundation for today's Grid software and ideas.

Others have been quick to realize the enormous potential of this idea and GRID projects have now been set up in many areas of the world. ESA has started an internal GRID initiative to look into the possibilities of using the GRID for a wide range of space applications and is also involved in international GRID projects such as DataGRID, which is funded by the European Union.

There are several other places to find out about Grids and given the increasing popularity and importance of this technology, additional literature will be added at a rapid rate. The first seminal book on Grids was edited by Foster and Kesselman (1999); the website "http://www.grid forum.org" is a link to the Global Grid Forum, an important community consortium for Grid activities, standardization, and best practices efforts. The website "http://www.globus.org" is the home page of the Globus project, which has had and continues to have a major influence on Grid Computing. A selection of web information sources for the Grid are as follows:

- Online Grid magazine published by EarthWeb, which is a division of INT Media Group: "http://www.gridcomputingplanet.com/";

- **G**rid **C**omputing **I**nformation **C**enter: "http://www.gridcomputing.com/";
- Informal listing of Grid News: "http://www.thegridreport.com/";
- Online Grid Magazine published by Tabor Griffin Communications: "http://www.gridtoday.com";
- Grid Research Integration Deployment and Support Center "http://www.gridscenter.org/".

The book by Berman *et al.* (2003) is a collection of separate chapters written by pioneers and leaders in Grid Computing. Different perspectives will inevitably lead to some inconstancy of views and notation but will provide a comprehensive look at the state-of-the-art and best practices for a wide variety of areas within Grid Computing. Most chapters were written especially for this book in the summer of 2002. A few chapters are reprints of seminal papers of technical and historical significance. The homepage of the book in "http://www.grid2002.org" will supplement the book with references, summaries, and notes.

Today, the Grid has gone global, with many worldwide collaborations between US, European, and Asia-Pacific researchers. Funding agencies, commercial vendors, academic researchers, and national centers and laboratories have come together to form a community of broad expertise with enormous commitment to building the Grid. Moreover, research in the related areas of networking, digital libraries, peer-to-peer computing, collaboratories, etc. are providing additional ideas relevant to the Grid.

5.2 Basics of Grid Computing

Grids are intrinsically distributed and heterogeneous but must be viewed by the user (whether an individual or another computer) as a virtual environment with uniform access to resources. Much of Grid software technology addresses the issues of resource scheduling, quality of service, fault tolerance, decentralized control and security, etc., which enable the Grid to be perceived as a single virtual platform by the user. For example, Grid security technologies must ensure that a single sign-on will generate security credentials that can be used for the many different actions potentially on multiple resources that may be needed in a Grid session. For some researchers and developers, the Grid is viewed as the future of the Web or Internet computing. The Web largely consists of individuals talking independently to servers; the Grid provides a collection of servers and clients often working collectively together to solve a problem. Computer-to-computer traffic is characteristic of Grids and both exploits and drives the increasing network backbone bandwidth.

The term "Grid" was originally used in analogy to other infrastructure (such as electrical power) grids. We will see in this book that the analogy correctly characterizes some aspects of Grid Computing (ubiquity for example) but not others (the performance variability of different resources which can support the same code, for example).

The Grid is the computing and data management infrastructure that will provide the electronic underpinning for a global society in business, government, research, science, and entertainment (Foster and Kesselman, 1999; Berman *et al.*, 2003). The Grid infrastructure will provide us with the ability to dynamically link together resources as an ensemble to support the execution of large-scale, resource-intensive, and distributed applications. Large-scale grids are intrinsically distributed, heterogeneous, and dynamic. They promise effectively infinite cycles and storage, as well as access to instruments, visualization devices, etc., without regard to geographic location (Berman *et al.*, 2003). The reality is that to achieve this promise, complex systems of software and services must be developed which allow access in a user-friendly way, which allow resources to be used together efficiently, and which enforce policies that allow communities of users to coordinate resources in a stable, perform promoting fashion. Whether users access the Grid to use one resource (a single computer, data archive, etc.) or to use several resources in aggregate as a coordinated "virtual computer", the Grid allows users to interface with the resources in a uniform way, providing a comprehensive and powerful platform for global computing and data management.

In the late 1990s, Grid researchers came together in the Grid Forum, evolving into the Global Grid Forum (GGF) (Abramson, 1995), where much of the early research is now evolving into the standards base for future Grids. Recently, the Global Grid Forum has been instrumental in the development of the Open Grid Services Architecture (OGSA), which integrates Globus and Web services approaches (Chapters 6–8). OGSA is being developed by both US and European initiatives aiming to define core services for a wide variety of areas including the following:

- Workload/Performance Management
- Security

- Availability/Service Management
- Logical Resource Management
- Clustering Services
- Connectivity Management
- Physical Resource Management
- Systems Management and Automation

5.2.1 *The nature of grid architecture*

The Grid has both a technology pull and an application push. Both these aspects and their integration (manifested in the activity of programming the Grid) are covered in the book written by Foster and Kesselman (Foster and Kesselman, 1999), which is divided into four parts. Chapters 1–5 (Part A) provide a basic motivation and overview of where we are today and how we got there. Part B, Chapters 6–19, covers the core Grid architecture and technologies. Part C, Chapters 20–34, describes how one can program the Grid and discusses Grid Computing Environments; these are covered both from the view of the user at a Web portal and of a computer program interacting with other Grid resources. Part D, Chapters 35–43, covers Grid applications, both ongoing and emerging applications representing the broad scope of programs that can be executed on Grids.

The real and specific problem that underlies the Grid concept is *coordinated resource sharing and problem solving in dynamic, multiinstitutional virtual organizations* (Foster *et al.*, 2001). The sharing that we are concerned with is not primarily file exchange but rather direct access to computers, software, data, and other resources, as is required by a range of collaborative problem-solving and resource brokering strategies emerging in industry, science, and engineering. This sharing is, necessarily, highly controlled, with resource providers and consumers defining clearly and carefully just what is shared, who is allowed to share, and the conditions under which sharing occurs. A set of individuals and/or institutions defined by such sharing rules form what we call a *virtual organization* (VO).

Foster *et al.* (2001) proposed a Grid architecture, which is not to provide a complete enumeration of all required protocols (and services, APIs, and SDKs) but rather to identify requirements for general classes of component. The result is an extensible, open architectural structure within which can be placed solutions to key VO requirements. Their architecture organizes components into layers. Components within each layer share common characteristics but can build on capabilities and behaviors provided by any lower layer.

In specifying the various layers of the Grid architecture, they followed the principles of the "hourglass model" (National Academy Press *et al.*, 1994). The narrow neck of the hourglass defines a small set of core abstractions and protocols (e.g., TCP and HTTP on the Internet), onto which many different high-level behaviors can be mapped (the top of the hourglass), and which themselves can be mapped onto many different underlying technologies (the base of the hourglass). By definition, the number of protocols defined at the neck must be small. In our architecture, the neck of the hourglass consists of *Resource* and *Connectivity* protocols, which facilitate the sharing of individual resources. Protocols at these layers are designed so that they can be implemented on top of a diverse range of resource types, defined at the *Fabric* layer, and can in turn be used to construct a wide range of global services and application-specific behaviors at the *Collective* layer — so called because they involve the coordinated ("collective") use of multiple resources.

5.2.1.1 *Fabric: Interfaces to local control*

The Grid *Fabric* layer provides the resources to which shared access is mediated by Grid protocols, for example, computational resources, storage systems, catalogs, network resources, and sensors. A "resource" may be a logical entity, such as a distributed file system, computer cluster, or distributed computer pool; in such cases, a resource implementation may involve internal protocols (e.g., the NFS storage access protocol or a cluster resource management system's process management protocol), but these are not the concern of Grid architecture.

The following brief and partial list provides a resource-specific characterization of capabilities (Foster *et al.*, 2001):

- *Computational resources*: Mechanisms are required for starting programs and for monitoring and controlling the execution of the resulting processes. Management mechanisms that allow control over the resources allocated to processes are useful, as are advance reservation mechanisms. Enquiry functions are needed for determining hardware and software characteristics as well as relevant state information such as current load and queue state in the case of scheduler-managed resources.

- *Storage resources*: Mechanisms are required for putting and getting files. Third-party and high-performance (e.g., striped) transfers are useful (Leigh *et al.*, 1997). So are mechanisms for reading and writing sub-sets of a file and/or executing remote data selection or reduction functions (Berman *et al.*, 1996). Management mechanisms that allow control over the resources allocated to data transfers (space, disk bandwidth, network bandwidth, and central processing unit (CPU)) are useful, as are advance reservation mechanisms. Enquiry functions are needed for determining hardware and software characteristics as well as relevant load information such as available space and bandwidth utilization.
- *Network resources*: Management mechanisms that provide control over the resources allocated to network transfers (e.g., prioritization and reservation) can be useful. Enquiry functions should be provided to determine network characteristics and load.
- *Code repositories*: This specialized form of storage resource requires mechanisms for managing versioned source and object code, for example, a control system such as CVS.
- *Catalogs*: This specialized form of storage resource requires mechanisms for implementing catalog queries and update operations, for example, a relational database (Barry *et al.*, 1998).

5.2.1.2 *Connectivity: Communicating easily and securely*

The *Connectivity* layer defines core communication and authentication protocols required for Grid-specific network transactions. Communication protocols enable the exchange of data between Fabric layer resources. Authentication protocols build on communication services to provide cryptographically secure mechanisms for verifying the identity of users and resources.

Communication requirements include transport, routing, and naming. While alternatives certainly exist, we assume here that these protocols are drawn from the TCP/IP protocol stack: specifically, the Internet (IP and ICMP), transport (TCP and UDP), and application (DNS, OSPF, and RSVP) layers of the Internet layered protocol architecture (AVAKI, 2001). This is not to say that in the future, Grid communications will not demand new protocols that take into account particular types of network dynamics.

5.2.1.3 *Resource: Sharing single resources*

The Resource layer builds on Connectivity layer communication and authentication protocols to define protocols (and APIs and SDKs) for the secure negotiation, initiation, monitoring, control, accounting, and payment of sharing operations on individual resources. Resource layer implementations of these protocols call Fabric layer functions to access and control local resources. Resource layer protocols are concerned entirely with individual resources and hence ignore issues of global state and atomic actions across distributed collections; such issues are the concern of the Collective layer discussed next.

Two primary classes of Resource layer protocols can be distinguished (Foster *et al.*, 2001):

- *Information protocols* are used to obtain information about the structure and state of a resource, for example, its configuration, current load, and usage policy (e.g., cost).
- *Management protocols* are used to negotiate access to a shared resource, specifying, for example, resource requirements (including advanced reservation and quality of service) and the operation(s) to be performed, such as process creation or data access. Since management protocols are responsible for instantiating sharing relationships, they must serve as a "policy application point," ensuring that the requested protocol operations are consistent with the policy under which the resource is to be shared. Issues that must be considered include accounting and payment. A protocol may also support monitoring the status of an operation and controlling (for example, terminating) the operation.

5.2.1.4 *Collective: Coordinating multiple resources*

While the Resource layer is focused on interactions with a single resource, the next layer in the architecture contains protocols and services (and APIs and SDKs) that are not associated with any one specific resource but rather are global in nature and capture interactions across collections of resources. For this reason, we refer to the next layer of the architecture as the *Collective* layer. Since *Collective* components build on the narrow Resource and Connectivity layer "neck" in the protocol hourglass, they can implement a wide variety of sharing behaviors without placing new requirements on the resources being shared.

5.2.1.5 *Applications*

The final layer in our Grid architecture comprises the user applications that operate within a VO environment. Applications are constructed in terms of, and by calling upon, services defined at any layer. At each layer, we have well-defined protocols that provide access to some useful service: resource management, data access, resource discovery, and so forth. At each layer, APIs may also be defined whose implementation (ideally provided by third-party SDKs) exchange protocol messages with the appropriate service(s) to perform desired actions.

Foster *et al.* (2001) described their grid architectural at high level and placed few constraints on design and implementation. To make this abstract discussion more concrete, they also list, for illustrative purposes, the protocols defined within the Globus Toolkit (Dinda and O'Hallaron, 1999), and used within such Grid projects as the NSF's National Technology Grid (Lee and Talia, 2003), NASA's Information Power Grid (Gabriel *et al.*, 1998), DOE's DISCOM (Baru *et al.*, 1998), GriPhyN (www.griphyn.org), NEESgrid (www.neesgrid.org), Particle Physics Data Grid (www.ppdg.net), and the European Data Grid (www.eu-datagrid.org).

5.2.2 *Grid computing tools*

Globus Toolkit: The Globus Toolkit has been designed to use (primarily) existing fabric components, including vendor-supplied protocols and interfaces. However, if a vendor does not provide the necessary Fabric-level behavior, the Globus Toolkit includes the missing functionality. For example, enquiry software is provided for discovering structure and state information for various common resource types, such as computers (e.g., OS version, hardware configuration, load (Czajkowski *et al.*, 1998), and scheduler queue status), storage systems (e.g., available space), and networks (e.g., current and predicted future load (Goux *et al.*, 2000; Linn, 2000)), and for packaging this information in a form that facilitates the implementation of higher-level protocols, specifically at the Resource layer. Resource management, on the other hand, is generally assumed to be the domain of local resource managers. One exception is the General-purpose Architecture for Reservation and Allocation (GARA) (Foster and Karonis, 1998), which provides a "slot manager" that can be used to implement advance reservation for resources that do not support this capability. Others have developed enhancements to the Portable Batch System (PBS) (Howell and Kotz, 2000) and Condor (Gasser and McDermott, 1990; Gong, 2002) that support advance reservation capabilities.

Grid programming must manage computing environments that are inherently parallel, distributed, heterogeneous, and dynamic, both in terms of the resources involved and their performance. Furthermore, grid applications will want to dynamically and flexibly compose resources and services across that dynamic environments. While it may be possible to build grid applications using established programming tools, they are not particularly well suited to effectively manage flexible composition or deal with heterogeneous hierarchies of machines, data, and networks with heterogeneous performance. Lee and Talia (2003) discussed issues, properties, and capabilities of grid programming models and tools to support efficient grid programs and their effective development. The main issues were outlined as portability, interoperability, and adaptivity, resource discovery, performance, fault tolerance, security, and program meta-models.

Lee and Talia (2003) provided a brief survey of many specific tools, languages, and environments for grids. Many, if not most, of these systems have their roots in "ordinary" parallel or distributed computing and are being applied in grid environments because they are established programming methodologies. Broader surveys are available in Lee *et al.* (2001) and Skillicorn and Talia (1998).

5.2.2.1 *Shared-state models*

Shared-state programming models are typically associated with tightly coupled, synchronous languages and execution models that are intended for shared-memory machines or distributed memory machines with a dedicated interconnection network that provides very high bandwidth and low latency.

JavaSpaces (Freeman *et al.*, 1999) is a Java-based implementation of the Linda tuple space concept, in which tuples are represented as serialized objects. The use of Java allows heterogeneous clients and servers to interoperate, regardless of their processor architectures and operating systems. The model used by JavaSpaces views an application as a collection of processes communicating between them by putting and getting objects into one or more *spaces.* A *space* is a shared and persistent object repository that is accessible via network. A programmer that wants to build a space-based application should design *distributed data structures* as a set of objects that are stored in one or more *spaces*. The new approach that the JavaSpaces programming model gives to the programmer makes building distributed applications much easier, even when dealing with such dynamic environments. Currently, efforts to implement JavaSpaces on grids using Java toolkits based on Globus are ongoing (Rana *et al*., 2002; Saelee and Rana, 2001).

5.2.2.2 *Message-passing models*

In message-passing models, processes run in disjoint address spaces and information are exchanged using message passing of one form or another.

The Message Passing Interface (MPI) (Message Passing Interface Forum, 1995, 1997) is a widely adopted standard that defines a two-sided message passing library, i.e., with matched sends and receives, that is well suited for grids. Many implementations and variants of MPI have been produced. The most prominent for grid computing is MPICH-G2.

MPICH-G2 (Foster and Karonis, 1998) is a grid-enabled implementation of the MPI that uses the Globus services (e.g., job startup and security) and allows programmers to couple multiple machines, potentially of different architectures, to run MPI applications. MPICH-G2 automatically converts data in messages sent between machines of different architectures and supports multiprotocol communication by automatically selecting TCP for inter-machine messaging and vendor-supplied MPI for intra-machine messaging.

5.2.2.3 *RPC and RMI models*

Remote Procedure Call (RPC) and Remote Method Invocation (RMI) models provide the same capabilities as those provided by MPI but structure the interaction between sender and receiver more as a language construct rather than a library function call that simply transfers an uninterpreted buffer of data between points A and B. RPC and RMI models provide a simple and well-understood mechanism for managing remote computations.

5.2.2.4 *Grid-enabled RPC*

GridRPC (Nakada, 2002) is an RPC model and API for grids. Besides providing standard RPC semantics with asynchronous, coarse-grain, task-parallel execution, it provides a convenient, high-level abstraction whereby the many details of interacting with a grid environment can be hidden. Three very important grid capabilities that GridRPC could transparently manage for the user are *Dynamic resource discovery and scheduling, Security, and Fault Tolerance.*

5.2.2.5 *Java RMI*

Java Remote Method In-vocation (RMI) enables a programmer to create distributed Java-based applications, in which the methods of remote Java objects can be invoked from other Java virtual machines, possibly on different hosts.

5.2.2.6 *Hybrid models*

The inherent nature of grid computing is to make all manner of hosts available to grid applications. Hence, some applications will want to run both within and across address spaces. That is to say, they will want to run perhaps multithreaded within a shared-address space and also by passing data and control between machines.

OpenMP (OpenMP Consortium, 1997) is a library that supports parallel programming in shared-memory parallel machines. It has been developed by a consortium of vendors with the goal of producing a standard programming interface for parallel shared-memory machines that can be used within mainstream languages, such as Fortran, C, and C++.

OmniRPC (Sato *et al.*, 2001) was specifically designed as a thread-safe RPC facility for clusters and grids. OmniRPC uses OpenMP to manage thread-parallel execution while using Globus to manage grid interactions.

All of these programming concepts can be put into one package, as is the case with message-passing Java, or MPJ (Carpenter *et al.*, 2000).

5.2.2.7 *Peer-to-peer models*

Peer-to-peer (P2P) computing (Gong, 2002) is the sharing of computer resources and services by direct exchange between systems. Peer-to-peer computing takes advantage of existing desktop computing power and networking connectivity, allowing economical clients to leverage their collective power to benefit the entire enterprise. In a peer-to-peer architecture, computers that have traditionally been used solely as clients communicate directly among themselves and can act as both clients and servers, assuming whatever role is most efficient for the network. This reduces the load on servers and allows them to perform specialized services (such as mail-list generation and billing) more effectively. A family of protocols specifically designed for peer-to-peer computing is JXTA (Gong, 2001). JXTA is a set of open, generalized peer-to-peer protocols, defined as XML messages, that allow any connected device on the network ranging from cell phones and wireless PDAs to PCs and servers to communicate and collaborate in a P2P manner.

5.2.2.8 *Frameworks, component models, and portals*

Besides these library and language tool approaches, entire programming environments facilitate the development and deployment of distributed applications. Lee and Talia (2002) broadly classified these approaches as frameworks, component models, and portals. A few important examples are reviewed here.

(a) Cactus: The Cactus Code and Computational Toolkit (Cactus Webmeister, 2000) provides a modular framework for computational physics. As a framework, Cactus provides application programmers with a high-level API for a set of services tailored for computational science. Besides support for services like parallel I/O and parallel checkpointing and restart, there are services for computational steering (dynamically changing parameters during a run) and remote visualization. To build a Cactus application, a user builds modules, called *thorns*, that are plugged into the framework *flesh*.

(b) CORBA: The Common Object Request Broker Architecture (CORBA) (The Object Management Group, 2000) is a standard tool where a meta-language interface is used to manage interoperability among objects. Object member access is defined using the Interface Definition Language (IDL). An Object Request Broker (ORB) is used to provide resource discovery among client objects. While CORBA can be considered middleware, its primary goal has been to manage interfaces between objects.

(c) CoG Kit: There are also efforts to make CORBA services directly available to grid computations. This is being done in the CoG Kit project (Verma *et al.*, 2001) to enable "Commodity Grids" through an interface layer that maps Globus services to a CORBA API.

(d) Legion: Legion (Lewis and Grimshaw, 1995) provides objects with a globally unique (and opaque) identifier. Using such an identifier, an object, and its members, can be referenced from anywhere. Being able to generate and dereference globally unique identifiers requires a significant distributed infrastructure. We note that all Legion development is now being done as part of the AVAKI Corporation (AVAKI, 2001).

(e) Component Architectures: Components extend the object-oriented paradigm by enabling objects to manage the interfaces they present and discover those presented by others (Szyperski, 1999). A number of component and component-like systems have been defined. These include COM/DCOM (Sessions, 1997), the CORBA 3 Component Model (The Object Management Group, 2000), Enterprise Java Beans and Jini (Englander, 1997; The Jini Community, 2000), and the Common Component Architecture (Gannon *et al.*, 2000).

(f) Portals: Portals can be viewed as providing a web-based interface to a distributed system. Commonly, portals entail a three-tier architecture that consists of (1) a first tier of clients, (2) a middle tier of brokers or servers, and (3) a third tier of object repositories, compute servers, databases, or any other resource or service needed by the portal. A number of examples are possible in this area. One is the Grid Portal Toolkit, aka GridPort (Thomas *et al.*, 2001). The GridPort Toolkit is partitioned into two parts: (1) the client interface tools and (2) the web portal services module. The client interface tools enable customized portal interface development and do not require users to have any specialized knowledge of the underlying portal technology.

Another very important example is the XCAT Science Portal (Krishnan *et al.*, 2001).

5.2.2.9 *Web service models*

Grid technologies are evolving toward an Open Grid Services Architecture (OGSA) (Foster *et al*., 2002) in which a grid provides an extensible set of services that virtual organizations can aggregate in various ways. OGSA defines a uniform exposed service semantics (the so-called grid service) based on concepts and technologies from both the Grid and Web services communities. OGSA defines standard mechanisms for creating, naming, and discovering transient grid service instances, provides location transparency and multiple protocol bindings for service instances, and supports integration with underlying native platform facilities.

5.2.2.10 *Coordination models*

The purpose of a coordination model is to provide a means of integrating a number of possibly heterogeneous components together, by interfacing with each component in such a way that the collective set forms a single application that can execute on parallel and distributed systems (Marinescu and Lee, 2002). Coordination models can be used to distinguish the computational concerns of a distributed or parallel application from the cooperation ones, allowing separate development but also the eventual fusion of these two development phases.

5.2.3 *Visualization*

As we enter the 21st century, the traditional model of stand-alone, centralized computer systems is rapidly evolving toward grid-based computing across a ubiquitous, continuous, and pervasive national computational infrastructure. While many application scenarios are not yet possible, grid computing will provide access to distributed resources with the same ease we now access electrical power.

Clearly, the ability to analyze data within this distributed environment is critical to the success of the grid. To successfully analyze data, we need visualization tools. The special issue of IEEE Computer Graphics and Applications (March/April 2003) provides a glimpse at the future of grid visualization by exploring working prototypes that exploit distributed resources.

5.2.4 *Grid applications and application middleware*

The Grid will serve as an enabling technology for a broad set of applications in science, business, entertainment, health, and other areas. However, the community faces a "chicken and egg" problem common to the development of new technologies: applications are needed to drive the research and development of the new technologies, but applications are difficult to develop in the absence of stable and mature technologies. In the Grid community, Grid infrastructure efforts, application development efforts, and middleware efforts have progressed together, often through the collaborations of multidisciplinary teams. In this section, we will discuss the successful Grid application and application middleware efforts related to geoinformation. As we continue to develop the software infrastructure that better realizes the potential of the Grid, and as common Grid infrastructure continues to evolve to provide a stable platform, the application and user community for the Grid will continue to expand.

Functionally, Grids are tools, middleware, and services for the following (Johnston, 1999):

- building the application frameworks that allow discipline scientists to express and manage the simulation, analysis, and data management aspects of overall problem solving;
- providing a uniform and secure access to a wide variety of distributed computing and data resources;
- supporting construction, management, and use of widely distributed application systems;
- facilitating human collaboration through common security services, and resource and data sharing;
- providing support for remote access to, and operation of, scientific and engineering instrumentation systems;
- managing and operating this computing and data infrastructure as a persistent service.

5.3 Grid Computing and Geoinformation Processing

Dozens of satellites are constantly collecting data about our planetary system — the Earth in particular — 24 hours a day, 365 days a year.

Satellite data are used for many purposes, for example, for telecommunications, navigation systems, and environmental monitoring. However, even with the most powerful computers processing, all these data are time-consuming and expensive. Distributing these tasks over a number of low-cost internet-connected platforms would provide enormous potential, at a relatively low cost, for many space applications.

5.3.1 *Spatial information grid*

The European DataGrid is a project funded by the European Union with the aim of setting up a computational and data-intensive grid of resources for the analysis of data coming from scientific exploration. Next-generation science will require co-ordinated resource sharing, collaborative processing, and analysis of huge amounts of data produced and stored by many scientific laboratories belonging to several institutions.

The main goal of the DataGrid initiative is to develop and test the technological infrastructure that will enable the implementation of scientific "collaboratories" where researchers and scientists will perform their activities regardless of geographical location. It will also allow interaction with colleagues from sites all over the world as well as the sharing of data and instruments on a scale previously unattempted. The project will devise and develop scalable software solutions and testbeds in order to handle many PetaBytes of distributed data, tens of thousands of computing resources (processors, disks, etc.) and thousands of simultaneous users from multiple research institutions.

5.3.2 *SpaceFRID*

GRID projects have enormous potential. They could provide scientists all over the world with easy access to unprecedented levels of computing resources and initiate a new era of science and scientific cooperation. ESA's SpaceGRID project is an important stepping stone in finding out how this potential can be used within Europe for space applications.

To look into this exciting possibility ESA started in September 2001 to work on its SpaceGRID study financed by the Agency's General Study Programme. This study is run by an international consortium of industry and research centers led by Datamat S.p.A. (Italy). Other members include Alcatel Space (France), CS Systemes d'Information (France),

Science Systems Plc (UK), QinetiQ (UK), and the Rutherford Appleton Laboratory of the UK Council for the Central Laboratory of the Research Councils.

The SpaceGRID study will carefully examine the middleware selected in the DataGRID project (Globus being one of the key elements) for use as well for the SpaceGRID study demonstrators. Two other aspects of this project are of particular importance for ESA: finding a way to ensure that the data processed by the SpaceGRID can be made available to public and educational establishments and ensuring that SpaceGRID activities are coordinated with other major international initiatives.

The project aims to assess how GRID technology can serve requirements across a large variety of space disciplines, sketch the design of an ESA-wide GRID infrastructure, foster collaboration, and enable shared efforts across space applications. It will analyze the highly complicated technical aspects of managing, accessing, exploiting, and distributing large amounts of data and set up test projects to see how well the GRID performs at carrying out specific tasks in EO, space weather, space science, and spacecraft engineering.

The goals of SpaceGRID are as follows:

(1) to assess how GRID technology can serve requirements across a large variety of space disciplines (spacecraft mechanical engineering, space weather, space science, and EO);
(2) to foster collaboration and enable shared efforts across space applications;
(3) to sketch the design of an ESA-wide (and common) infrastructure;
(4) to demonstrate proof of concept through prototyping;
(5) to involve both industry and research centers;
(6) keep Europe up with GRID efforts!

The involved domains include EO, Space Research — Space Weather, Space Research — Spacecraft Plasma Interaction, Space Research — Radiation Transport, Solar System Research, and Spacecraft Engineering. Figure 5.1 shows the working plan of the SpaceGRID project.

For the EO domain, at present, several thousands of ESA users worldwide have online access to EO missions related to meta-data (10 million

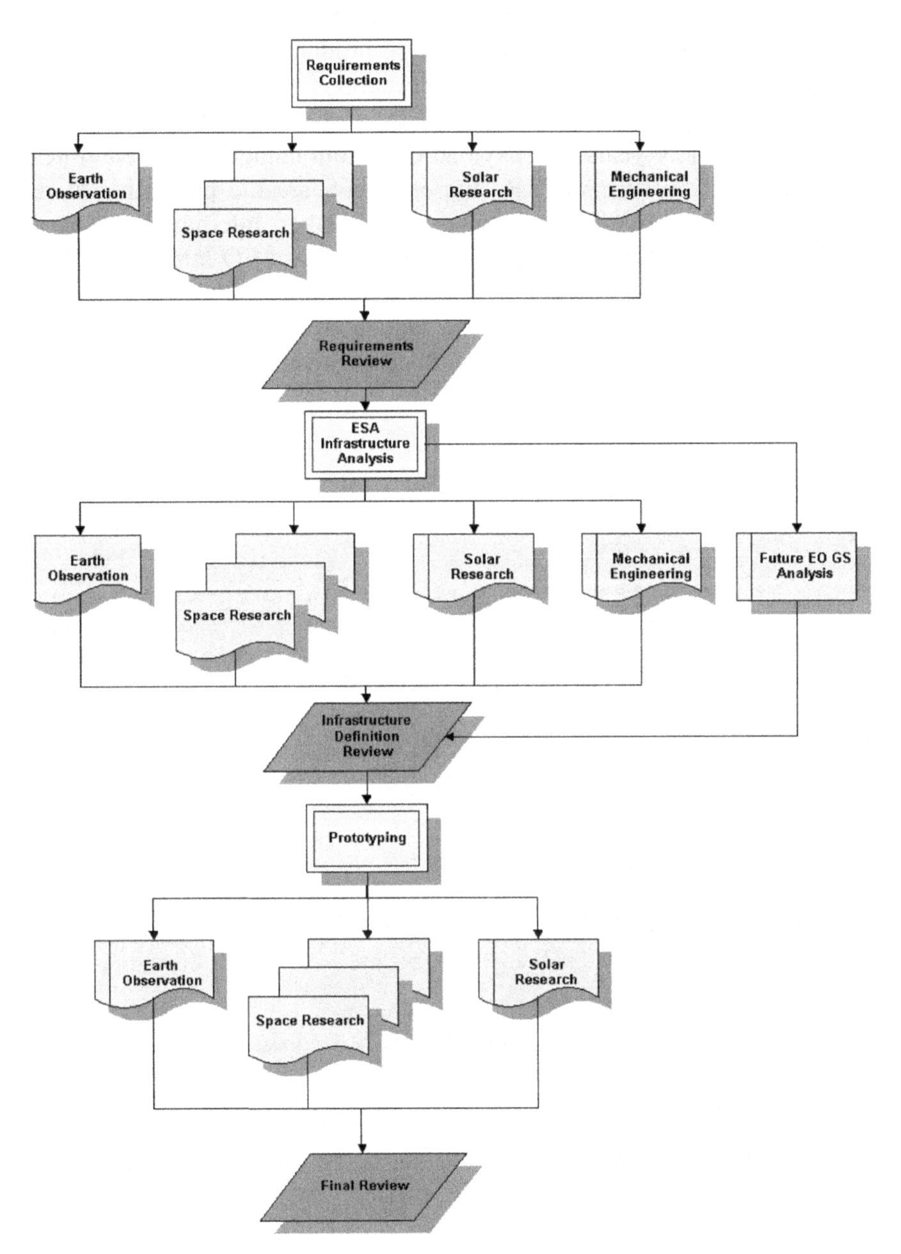

Figure 5.1. SpaceGRID working plan.

references), data, and derived information products acquired, processed, and archived by more than 30 worldwide stations. ESA-ESRIN is the core management facility of the European EO infrastructure which operates, since more than 20 years, EO payload data from numerous EO Satellites (ERS, Landsat, JERS, AVHRR, and SeaWiFS are the presently active missions) owned by ESA, European, and other International Space Agencies (see Figure 5.2). Currently, ESA-managed EO missions download about 100 GBytes of data per day. This number will grow up to 500 GBytes in the near future, after the launch of the next ESA mission, ENVISAT, due autumn 2001. The products archived by ESA facilities are

Figure 5.2. European EO infrastructure.

estimated at present at 800 Terabytes and will grow in the future with a rate of at least 500 Terabytes per year.

When dealing with EO applications, we have always to face two different aspects: the capability to have access to distributed datasets and the specificity of every single application.

That is why the first task is the requirements definition that will analyze EO domain from both these viewpoints.

Concerning the distributed data, together with the definition of requirements for seamless access, it is also required to identify and define requirements for EO metadata and data dictionaries, based on existing standards. This activity aims to define a common way to specify and use EO data in a GRID environment.

At the same time, it is required to define the application requirements for such an environment, also in terms of performances and resources, the following step being the identification of application(s) matching the above requirements to be tested over the SpaceGRID testbed.

The choice of application(s) will be performed jointly with the ESA Technical Officer, focusing on that (those) significant enough to demonstrate how the EO will benefit from the GRID technology.

Last, but not least, requirements for collaborative environments in the field of EO will have to be defined, with the aim of laying down the bases for the tools for the future of EO researchers/scientists' collaboration on common problems using common and shared datasets (note that the "vision" of today's GRID guru's comprehends also "Colleagues" among the GRID shareable resources, and tools supporting collaborative environment are the key for this!).

Still talking about analyzes in EO field, the study will envisage solutions for future EO ground segments based on GRID infrastructures, starting from an analysis of those that would be the requirements for future EO ground segments and identifying those parts that would take more advantages from a GRID-aware environment.

After the analyzes and requirements phase, the study will target the testbed and the prototypes.

Concerning the analysis of the ESA GRID Infrastructure to be used as testbed, the EO applications-related section will focus on specific topics, like distributed data and archives management, as described in the EO application requirements. Then, the prototyping activities will take place. The output will be a (set of) working prototype(s), to demonstrate solutions for seamless access to distributed EO data and resources. As above stated, the application(s) to be prototyped will be chosen jointly by the contractor and the Agency and will be possibly installed on a testbed that involves both ESRIN and ESTEC sites.

References

Abramson, D., Sosic, R., Giddy, J., and Hall, B. Nimrod. (1995). A tool for performing parameterized simulations using distributed workstations. In *Proceedings of 4th IEEE Symposium on High Performance Distributed Computing.*

Aiken, R., Carey, M., Carpenter, B., Foster, I., Lynch, C., Mambretti, J., Moore, R., Strasnner, J., and Teitelbaum, B. (2000). Network Policy and Services: A Report of a Workshop on Middleware, IETF, RFC 2768. http://www.ietf.org/rfc/rfc2768.txt.

Allcock, B., Bester, J., Bresnahan, J., Chervenak, A.L., Foster, I., Kesselman, C., Meder, S., Nefedova, V., Quesnel, D., and Tuecke, S. (2001). Secure, efficient data transport and replica management for high-performance data-intensive computing. In *Mass Storage Conference.*

Armstrong, R., Gannon, D., Geist, A., Keahey, K., Kohn, S., McInnes, L., and Parker, S. (1999). Toward a common component architecture for high performance scientific computing. In *Proceedings of 8th IEEE Symposium on High Performance Distributed Computing.*

Arnold, K., O'Sullivan, B., Scheifler, R. W., Waldo, J., and Wollrath, A. (1999). *The Jini Specification.* Boston: Addison-Wesley. See also www.sun.com/jini.

AVAKI. (2001). The AVAKI corporation. www.avaki.com.

Baker, F. (1995). Requirements for IP Version 4 Routers, IETF, RFC 1812. http://www.ietf.org/rfc/rfc1812.txt.

Barry, J., Aparicio, M., Durniak, T., Herman, P., Karuturi, J., Woods, C., Gilman, C., Ramnath, R., and Lam, H. (1998). NIIIP-SMART: An investigation of distributed object approaches to support MES development and deployment in a virtual enterprise. In *2nd International Enterprise Distributed Computing Workshop*, IEEE Press.

Baru, C., Moore, R., Rajasekar, A., and Wan, M. (1998). The SDSC storage resource broker. In *Proceedings of CASCON'98 Conference.*

Beiriger, J., Johnson, W., Bivens, H., Humphreys, S., and Rhea, R. (2000). Constructing the ASCI grid. In *Proceedings of 9th IEEE Symposium on High Performance Distributed Computing, 2000.* IEEE Press.

Benger, W., Foster, I., Novotny, J., Seidel, E., Shalf, J., Smith, W., and Walker, P. (1999). Numerical relativity in a distributed environment. In *Proceedings of 9th SIAM Conference on Parallel Processing for Scientific Computing.*

Berman, F., Fox, G., and Hey, T. (2003). *Grid Computing: Making the Global Infrastructure a Reality*. Chichester: John Wiley & Sons Ltd.

Berman, F. (1999). High-performance schedulers. In Foster, I. and Kesselman, C. (Eds.), *The Grid: Blueprint for a New Computing Infrastructure*, pp. 279–309. Burlington: Morgan Kaufmann.

Berman, F., Wolski, R., Figueira, S., Schopf, J., and Shao, G. (1996). Application-level scheduling on distributed heterogeneous networks. In *Proceedings of Supercomputing'96*.

Beynon, M., Ferreira, R., Kurc, T., Sussman, A., and Saltz, J. (2000). DataCutter: Middleware for filtering very large scientific datasets on archival storage systems. In *Proceedings of 8th Goddard Conference on Mass Storage Systems and Technologies/17th IEEE Symposium on Mass Storage Systems*, pp. 119–133.

Bolcer, G. A. and Kaiser, G. (1999). SWAP: Leveraging the web to manage workflow. *IEEE Internet Computing*, 85–88.

Brunett, S., Czajkowski, K., Fitzgerald, S., Foster, I., Johnson, A., Kesselman, C., Leigh, J., and Tuecke, S. (1998). Application experiences with the globus toolkit. In *Proceedings of 7th IEEE Symposium on High Performance Distributed Computing*, pp. 81–89. IEEE Press.

Butler, R., Engert, D., Foster, I., Kesselman, C., Tuecke, S., Volmer, J., and Welch, V. (2000). Design and deployment of a national-scale authentication infrastructure. *IEEE Computer*, **33**(12): 60–66.

Cactus Webmeister. (2000). The Cactus Code Website. www.CactusCode.org.

Camarinha-Matos, L. M., Afsarmanesh, H., Garita, C., and Lima, C. Towards an architecture for virtual enterprises. *Journal of Intelligent Manufacturing*.

Carpenter, B., *et al.* (2000). MPJ: MPI-like message-passing for Java. *Concurrency: Practice and Experience*, **12**(11): 1019–1038.

Casanova, H. and Dongarra, J. (1997). NetSolve: A network server for solving computational science problems. *International Journal of Supercomputer Applications and High Performance Computing*, **11**(3): 212–223.

Casanova, H., Dongarra, J., Johnson, C., and Miller, M. (1999). Application-specific tools. In Foster, I. and Kesselman, C. (Eds.), *The Grid: Blueprint for a New Computing Infrastructure*, pp. 159–180. Burlington: Morgan Kaufmann.

Casanova, H., Obertelli, G., Berman, F., and Wolski, R. (2000). The AppLeS parameter sweep template: User-level middleware for the grid. In *Proceedings of SC'2000*.

Chervenak, A., Foster, I., Kesselman, C., Salisbury, C., and Tuecke, S. (2001). The data grid: Towards an architecture for the distributed management and analysis of large scientific data sets. *Journal of Network and Computer Applications*.

Childers, L., Disz, T., Olson, R., Papka, M. E., Stevens, R., and Udeshi, T. (2000). Access grid: Immersive group-to-group collaborative visualization.

In *Proceedings of 4th International Immersive Projection Technology Workshop.*

Clarke, I., Sandberg, O., Wiley, B., and Hong, T.W. (1999). Freenet: A distributed anonymous information storage and retrieval system. In *ICSI Workshop on Design Issues in Anonymity and Unobservability.*

Czajkowski, K., Fitzgerald, S., Foster, I., and Kesselman, C. (2001). Grid information services for distributed resource sharing.

Czajkowski, K., Foster, I., and Kesselman, C. (1999). Co-allocation services for computational grids. In *Proceedings of 8th IEEE Symposium on High Performance Distributed Computing.* IEEE Press.

Czajkowski, K., Foster, I., Karonis, N., Kesselman, C., Martin, S., Smith, W., and Tuecke, S. (1998). A resource management architecture for metacomputing systems. In *The 4th Workshop on Job Scheduling Strategies for Parallel Processing*, pp. 62–82.

DeFanti, T., Foster, I., Papka, M., Stevens, R., and Kuhfuss, T. (1996). Overview of the I-WAY: Wide area visual supercomputing. *International Journal of Supercomputer Applications*, **10**(2): 123–130.

DeFanti, T. and Stevens, R. (1999). Teleimmersion. In Foster, I., and Kesselman, C. (Eds.), *The Grid: Blueprint for a New Computing Infrastructure*, pp. 131–155. Burlington: Morgan Kaufmann.

Dierks, T. and Allen, C. (1999). The TLS Protocol Version 1.0, IETF, RFC 2246. http://www.ietf.org/rfc/rfc2246.txt.

Dinda, P. and O'Hallaron, D. (1999). An evaluation of linear models for host load prediction. In *Proceedings of 8th IEEE Symposium on High-Performance Distributed Computing.* IEEE Press.

Englander, R. (1997). *Developing Java Beans.* Sebastopol: O'Reilly.

Foster, I. (2000). Internet computing and the emerging grid. *Nature Web Matters.* http://www.nature.com/nature/webmatters/grid/grid.html.

Foster, I. and Karonis, N. T. (1998). A grid-enabled MPI: Message passing in heterogeneous distributed computing systems. In *Supercomputing.* IEEE, November 1998. www.supercomp.org/sc98.

Foster, I. and Kesselman, C. (1998). The globus project: A status report. In *Proceedings of Heterogeneous Computing Workshop*, pp. 4–18. IEEE Press.

Foster, I. and Kesselman, C. (Eds.). (1999). *The Grid: Blueprint for a New Computing Infrastructure.* Burlington: Morgan Kaufmann.

Foster, I., Geisler, J., Nickless, W., Smith, W., and Tuecke, S. (1997). Software infrastructure for the I-WAY high performance distributed computing experiment. In *Proceedings of 5th IEEE Symposium on High Performance Distributed Computing*, pp. 562–571.

Foster, I., Kesselman, C., and Tuecke, S. (2001). The anatomy of the grid: Enabling scalable virtual organizations. *International Journal of Supercomputer Applications.*

Foster, I., Kesselman, C., Nick, J., and Tuecke, S. (2002, June). Grid services for distributed system integration. *IEEE Computer*, 37–46.

Foster, I., Kesselman, C., Tsudik, G., and Tuecke, S. (1998). A security architecture for computational grids. In *ACM Conference on Computers and Security*, pp. 83–91.

Foster, I., Roy, A., and Sander, V. (2000). A quality of service architecture that combines resource reservation and application adaptation. In *Proceedings of 8th International Workshop on Quality of Service*.

Freeman, E., Hupfer, S. and Arnold, K. (1999). *JavaSpaces: Principles, Patterns, and Practice*. Boston: Addison-Wesley.

Frey, J., Foster, I., Livny, M., Tannenbaum, T., and Tuecke, S. (2001). *Condor-G: A Computation Management Agent for Multi-institutional Grids*. University of Wisconsin Madison.

Gabriel, E., Resch, M., Beisel, T., and Keller, R. (1998). Distributed computing in a heterogeneous computing environment. In *Proceedings of EuroPVM/MPI'98*.

Gannon, D. and Grimshaw, A. (1999). Object-based approaches. In Foster, I., and Kesselman, C. (Eds.), *The Grid: Blueprint for a New Computing Infrastructure*, pp. 205–236. Burlington: Morgan Kaufmann.

Gannon, D., *et al.* (2000). CCAT: The Common Component Architecture Toolkit. www.extreme.indiana.edu/ccat.

Gasser, M. and McDermott, E. (1990). An architecture for practical delegation in a distributed system. In *Proceedings of 1990 IEEE Symposium on Research in Security and Privacy*, pp. 20–30. IEEE Press.

Gehring, J. and Reinefeld, A. (1996, May). MARS — A framework for minimizing the job execution time in a metacomputing environment. *Future Generation Computer Systems*, **12**(1): 87–99. https://doi.org/10.1016/0167-739X(95)00037-S.

Gong, L. (Ed.). (2002). *IEEE Internet Computing*, **6**(1), Special Issue on Peer-to-Peer Networking. IEEE.

Gong, L. (2001). JXTA: A network programming environment. *IEEE Internet Computing*, **5**(3).

Goux, J.-P., Kulkarni, S., Linderoth, J., and Yoder, M. (2000). An enabling framework for master-worker applications on the computational grid. In *Proceedings of 9th IEEE Symposium on High Performance Distributed Computing*. IEEE Press.

Grimshaw, A. and Wulf, W. (1996). Legion — A view from 50,000 feet. In *Proceedings of 5th IEEE Symposium on High Performance Distributed Computing*, pp. 89–99. IEEE Press.

Gropp, W., Lusk, E., and Skjellum, A. (1994). *Using MPI: Portable Parallel Programming with the Message Passing Interface*. Cambridge: MIT Press.

Hoschek, W., Jaen-Martinez, J., Samar, A., Stockinger, H. and Stockinger, K. (2000). Data management in an international data grid project. In *Proceedings of 1st IEEE/ACM International Workshop on Grid Computing*. Springer Verlag Press.

Howell, J. and Kotz, D. (2000). End-to-end authorization. In *Proceedings of 2000 Symposium on Operating Systems Design and Implementation*. USENIX Association.

Johnston, W. E., Gannon, D. and Nitzberg, B. (1999). Grids as production computing environments: The engineering aspects of NASA's information power grid. In *Proceedings of 8th IEEE Symposium on High Performance Distributed Computing*. IEEE Press.

Krishnan, S., *et al.* (2001). The XCAT science portal. In *Supercomputing*, November 2001.

Lee, C. and Talia, D. (2003, May). *Grid Programming Models: Current Tools, Issues and Directions*.

Lee, C., Matsuoka, S., Talia, D., Sussman, A., Mueller, M., Allen, G., and Saltz. J. (2001). A grid programming primer. http://www.gridforum.org/7APM/APS.htm. Submitted to the Global Grid Forum, August 2001.

Leigh, J., Johnson, A. and DeFanti, T. A. (1997). CAVERN: A distributed architecture for supporting scalable persistence and interoperability in collaborative virtual environments. *Virtual Reality: Research, Development and Applications*, **2**(2): 217–237.

Lewis, M. and Grimshaw, A. (1995). The core legion object model. Technical report, University of Virginia. TR CS-95-35.

Linn, J. (2000). Generic security service application program interface version 2, update 1, IETF, RFC 2743. http://www.ietf.org/rfc/rfc2743.

Litzkow, M., Livny, M. and Mutka, M. (1988). Condor — A hunter of idle workstations. In *Proceedings of 8th International Conference on Distributed Computing Systems*, 1988, pp. 104–111.

Livny, M. (1999). High-throughput resource management. In Foster, I., and Kesselman, C. (Eds.), *The Grid: Blueprint for a New Computing Infrastructure*, pp. 311–337. Burlington: Morgan Kaufmann.

Lopez, I., Follen, G., Gutierrez, R., Foster, I., Ginsburg, B., Larsson, O., S. Martin and Tuecke, S. (2000). NPSS on NASA's IPG: Using CORBA and globus to coordinate multidisciplinary aeroscience applications. In *Proceedings of NASA HPCC/CAS Workshop*. NASA Ames Research Center.

Lowekamp, B., Miller, N., Sutherland, D., Gross, T., Steenkiste, P. and Subhlok, J. (1998). A resource query interface for network-aware applications. In *Proceedings of 7th IEEE Symposium on High-Performance Distributed Computing*. IEEE Press.

Marinescu, D. and Lee, C. (Eds.). (2002). *Process Coordination and Ubiquitous Computing*. Boca Raton: CRC Press.

Message Passing Interface Forum. (1995). MPI: A message passing interface standard. www.mpi-forum.org/. June 1995.

Message Passing Interface Forum. (1997). MPI-2: Extensions to the Message Passing Interface. www.mpi-forum.org/. July 1997.

Moore, R., Baru, C., Marciano, R., Rajasekar, A., and Wan, M. (1999). Data-intensive computing. In Foster, I. and Kesselman, C. (Eds.), *The Grid: Blueprint for a New Computing Infrastructure*, pp. 105–129. Burlington: Morgan Kaufmann.

Nakada, H., Matsuoka, S., Seymour, K., Dongarra, J., Lee, C., and Casanova, H. (2002). GridRPC: A remote procedure call API for grid computing. Technical report, University of Tennesse, June 2002. ICL-UT-02-06.

Nakada, H., Sato, M., and Sekiguchi, S. (1999). Design and implementations of ninf: Towards a global computing infrastructure. *Future Generation Computing Systems*.

Novotny, J., Tuecke, S., and Welch, V. (2001). Initial Experiences with an Online Certificate Repository for the Grid: MyProxy.

OpenMP Consortium. (1997). *OpenMP C and C++ Application Program Interface, Version 1.0*.

Papakhian, M. (1998). Comparing job-management systems: The user's perspective. *IEEE Computational Science & Engineering*, April-June 1998. http://pbs.mrj.com.

Rana, O. F., Getov, V. S., Sharakan, E., Newhouse, S., and Allan, R. (2002). Building Grid Services with Jini and JXTA, February 2002. GGF2 Working Document.

Saelee, D. and Rana, O. F. (2001). Implementing services in a computational grid with jini and globus. In *1st EuroGlobus Workshop*. See www.euroglobus.unile.it.

Sato, M., Hirono, M., Tanaka, Y., and Sekiguchi, S. (2001). OmniRPC: A grid RPC facility for cluster and global computing in OpenMP. In *WOMPAT*. LNCS, Vol. 2104, pp. 130–136. Berlin: Springer-Verlag.

Sculley, A. and Woods, W. (2000). *B2B Exchanges: The Killer Application in the Business-to-Business Internet Revolution*. ISI Publications.

Sessions, R. (1997). *COM and DCOM: Microsoft's Vision for Distributed Objects*. Hoboken: John Wiley & Sons.

Skillicorn, D. and Talia, D. (1998). Models and languages for parallel computation. *ACM Computing Surveys*, **30**(2). June 1998.

Steiner, J., Neuman, B.C., and Schiller, J. (1988). Kerberos: An authentication system for open network systems. In *Proceedings of Usenix Conference*, pp. 191–202.

Stevens, R., Woodward, P., DeFanti, T., and Catlett, C. (1997). From the I-WAY to the national technology grid. *Communications of the ACM*, **40**(11): 50–61.

Szyperski, C. (1999). *Component Software: Beyond Object-oriented Programming*. Boston: Addison-Wesley.

The Jini Community. (2000). The Community Resource for Jini Technology. www.jini.org.

The Object Management Group. (2000). CORBA 3 Release Information. http://www.omg.org/technology/corba/corba3releaseinfo.htm.

Thomas, M., *et al*. (2001). The Grid Portal Toolkit. grid-port.npaci.edu.

Thompson, M., Johnston, W., Mudumbai, S., Hoo, G., Jackson, K., and Essiari, A. (1999). Certificate-based access control for widely distributed resources. In Proceedings of 8th Usenix Security Symposium, 1999.

Tierney, B., Johnston, W., Lee, J., and Hoo, G. (1996). Performance analysis in high-speed wide area IP over ATM networks: Top-to-bottom end-to-end monitoring. *IEEE Networking*.

Vahdat, A., Belani, E., Eastham, P., Yoshikawa, C., Anderson, T., Culler, D., and Dahlin, M. (1998). WebOS: Operating system services for wide area applications. In *7th Symposium on High Performance Distributed Computing*, July 1998.

Verma, S., Gawor, J., von Laszewski, G., and Parashar, M. (2001). A CORBA commodity grid kit. In Grid 2001, November 2001.

Weissman, J. B. (1999). Prophet: Automated scheduling of SPMD programs in workstation networks. *Concurrency: Practice and Experience*, **11**: 301–321.

Wolski, R. (1997). Forecasting network performance to support dynamic scheduling using the network weather service. In *Proceedings of 6th IEEE Symposium on High Performance Distributed Computing*, Portland, Oregon.

Web site associated with book

1. "Grid Computing: Making the Global Infrastructure a Reality". http://www.grid2002.org.
2. Global Grid Forum Web Site, http://www.gridforum.org.
3. Globus Project Web Site, http://www.globus.org.
4. Online Grid magazine published by EarthWeb, which is a division of INT Media Group, http://www.gridcomputingplanet.com/.
5. The Grid Computing Information Center, http://www.gridcomputing.com/.
6. Informal Listing of Grid News, http://www.thegridreport.com/.
7. Online Grid Magazine published by Tabor Griffin Communications, http://www.gridtoday.com.
8. Grid Research Integration Deployment and Support Center, http://www.gridscenter.org/.
9. The SCxy Series of Conferences, http://www.supercomp.org.

10. Gnutella Peer-to-peer (file-sharing) Client http://www.gnutella.com/ or http://www.gnutellanews.com/.
11. SETI@Home Internet Computing, http://setiathome.ssl.berkeley.edu/.
12. *Realizing the Information Future: The Internet and Beyond.* Washington, DC: National Academy Press, 1994. http://www.nap.edu/readingroom/books/rtif/.
13. The Network Weather Service, http://nws.cs.ucsb.edu/.
14. SDSC Storage Resource Broker, http://www.npaci.edu/DICE/SRB/.
15. Netsolve RPC Based Networked Computing, http://icl.cs.utk.edu/netsolve/.
16. Ninf Global Computing Infrastructure, http://ninf.apgrid.org/.

Chapter 6

High-Performance Computing for Remote Sensing Big Data

6.1 Remote Sensing Big Data

McKinsey, the world's leading consulting firm, was the first to suggest that the era of "Big Data" was upon us, saying, "Data, which has permeated every industry and business function today, has become an important factor in production. The mining and use of massive amounts of data heralds a new wave of productivity growth and consumer surplus."

An article published by the International Data Corporation (IDC) stated the following: The amount of data generated, copied, and accessed worldwide will double every two years. In 2020, approximately 59 zettabytes (ZB) of data will be generated, copied, and accessed worldwide. IDC estimates that this number will reach 149 ZB by 2024. The characteristics of Big Data are shown in Table 6.1.

Spatial Big Data refers to Big Data with or implying geographic location information, and Spatial Big Data has two major characteristics: first, it has or implies spatial location information; second, it has the characteristics of Big Data, and the characteristics of spatial Big Data can be expressed by "L+5V", namely Location, Volume, Velocity, Variety, Veracity, and Value (Zhong *et al.*, 2020). Some scholars proposed that in addition to the above features, spatial big data also has Visualization and Visibility features (Li *et al.*, 2016).

As a special type of Spatial Big Data, Remote Sensing Big Data refers to the theory, method, technology, and activity of obtaining industry-valued information from a massive remote sensing dataset by using big

Table 6.1. Big data characteristics.

Proposer	Characteristics	Content
Doug Laney	Volume	Volume of data
	Velocity	Speed of input and output data
	Variety	Source and type of data
IBM and Microsoft	Veracity	Clutter and trustworthiness of data
McKinsey	Value	The value of hidden insights in big data

data thinking and means, mainly based on a massive remote sensing dataset and integrating auxiliary data from other multiple sources. In addition to the characteristics of Big Data and Spatial Big Data, Remote Sensing Big Data has its own more specific characteristics, as shown in Table 6.2.

Table 6.2. Remote sensing big data characteristics.

Features	Specific characteristics
Large volume	The volume of remote sensing data is expanding continuously due to the accumulation of existing sensor information and the application of new sensors.
Diversity	The remote sensing data comes from different sensors in the sky, air, earth, and sea, and the types of data are characterized by diversity.
Multitemporal	Data acquired at different times, and data of long time series.
Multidimensional	Characterization of the same earth surface object in different dimensions.
Multiresolution	Data with different spatial and temporal resolutions, reflecting different granularity of information.
High value	A large amount of valuable information and knowledge can be mined from remote sensing data.

Nowadays, our ability to acquire remote sensing data has increased to an unprecedented level and we have entered an era of big data, where remote sensing data is clearly showing the characteristics of big data and remote sensing big data is getting more and more attention from government projects, commercial applications, and academic fields (Liu *et al.*, 2018). Remote sensing technology has been successfully used in several applications, such as agricultural applications (food security monitoring, rangeland monitoring, etc.), marine applications (ship detection, oil spill detection, etc.), urban planning, urban monitoring, human settlements (urban and rural), food security monitoring, water quality monitoring, energy assessment, disease population, ecosystem assessment, global warming, global change, global forest resource assessment, the ancient site discovery (archaeology), etc. (Chi *et al.*, 2016).

Advanced sensor technologies are revolutionizing the way remote sensing data are collected, managed, and analyzed. The explosive growth of remote sensing information is rapidly bringing new processing challenges, and the computational power required for remote sensing applications is greatly increased (Xue *et al.*, 2008). Global regional and long time series remote sensing data processing and analysis, near-real-time response requirements based on remote sensing data such as geological disasters, oil spills, floods, fires, military target surveillance, weather monitoring, etc., all need to be realized with efficient computing power. With the continuous development and application of remote sensing big data, the importance of High-Performance Computing technology becomes more and more prominent.

6.2 High-Performance Computing

High-Performance Computing (HPC) is a multidisciplinary field that combines hardware technologies and architectures, operating systems, programming tools, software, and end-user problems and algorithms, and a constant theme throughout HPC is "parallelism", which reduces task time by executing multiple operations simultaneously (Sterling *et al.*, 2017). In the 1970s, Michael Flynn proposed four parallel computing structures, SISD, SIMD, MIMD, and MISD, based on the parallelism between Data stream (D) and Instruction stream (I) (Single, S; Multiple, M), which simplified the classification of parallel systems and control methods, and the classification of control methods, which is still used by many scholars. Some scholars also classify massively parallel systems into Single Instruction Multiple Data Stream (SIMD) computers, Parallel Vector Processor (PVP), Symmetric Multiprocessor (SMP), Massively Parallel Processor (MPP), Cluster of Workstations (COW), and Distributed Shared Memory (DSM), the last five of which are Multi-Instruction Multiple Data Stream (MIMD) computers (Chen, 2011).

High-Performance Computing technology has gone through different stages of development; the early stage is mainly supercomputing, mainly through PVP, SMP, MPP, COW, and other development stages, as shown in Figure 6.1.

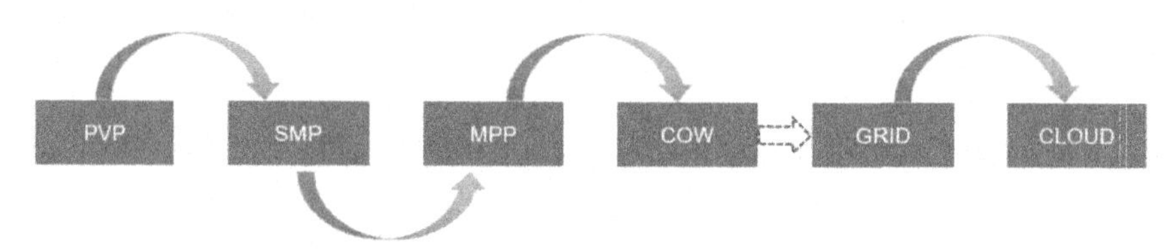

Figure 6.1. Schematic diagram of the development stages of HPC technology.

Parallel Vector Processing (PVP), by increasing the number of processors and expanding the memory, has occupied the HPC market for 20 years. In the early 1990s, Massively Parallel Processing (MPP) became the direction of HPC development. Clustering is an inexpensive and convenient method of HPC that connects multiple computers over a local network to accomplish work together. A grid integrates hundreds of millions of computers, storage, valuable devices, databases, etc., distributed on the Internet to form a virtual, unprecedentedly powerful supercomputer. Cloud computing is a computing model driven by the demand for big data storage analysis and resource elasticity expansion and contraction. It provides users with highly available, efficient, and elastic computing and storage resources and data functional services through a virtualized, dynamic, and scaled resource pool.

6.3 Supercomputers

Supercomputers, also known as high-performance computers, specifically refer to a class of computers with supercomputing power that are used to solve today's very large, very high, and very complex computing tasks (Zhou, 2011).

The world's first electronic digital computer, the ENIAC, was introduced in 1946. The development of high-performance computers can be traced back to the 1960s and 1970s when these computers were called "Mainframe". In the 1970s and 1980s, the "Supercomputer" appeared. In November 1975, ILILAC-IV was connected to ARPANET for distributed use and became the first available supercomputer, almost a year ahead of Cray-1. Since the 1990s, a new type of massively parallel supercomputer has emerged called the "High-Performance Computers, HPC". The development of high-performance computers is illustrated in Figure 6.2.

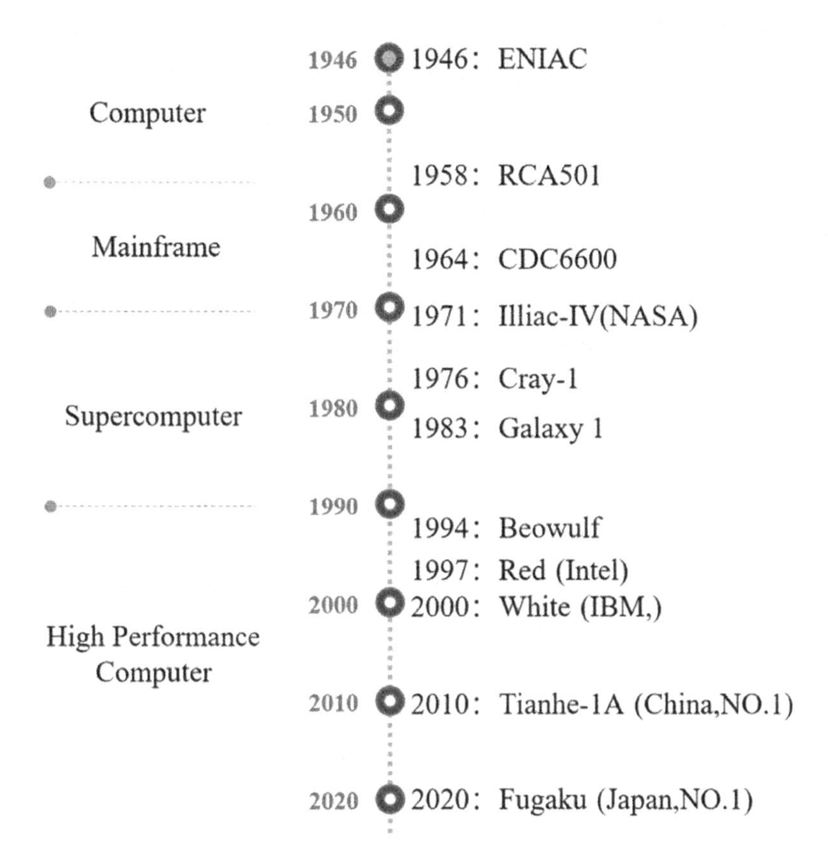

Figure 6.2. Development of high-performance computers.

Many scholars have used supercomputers to solve the problems that are difficult to be realized by traditional computing methods in Remote Sensing Big Data global surface classification and ecological environment simulation and have conducted research on many aspects of supercomputers for knowledge extraction, feature identification, and crop classification in the field of remote sensing big data. In earlier years, scholars' related research works based on supercomputers in remote sensing data processing are shown in Table 6.3.

The pattern of performance improvement of Supercomputers is basically consistent with Moore's law. The performance of supercomputers keeps increasing while the number of supercomputers keeps increasing. Table 6.4 shows the World Ranking of supercomputers in November 2021.

Table 6.3. Some relevant studies in the 1990s.

Scholars/Time	Supercomputers	Research
Lee *et al.* (1990)	Cray-2	Ocean observation satellite (MOS) data were processed to estimate the sea surface temperature of the Yellow Sea.
Yang *et al.* (1993)	Cray-2s	Realizes common functions such as preprocessing, image enhancement, image conversion, classification image compression, etc., applied to image GCP correction.
Armstlon *et al.* (1994)	KSR1	Inverse distance weighted interpolation algorithm.

Table 6.4. Global supercomputers top 10 (November 2021).

Rank	Name	Performance	Country
No. 1	Fugaku	Cores: 7630848 Rmax: 442010 TFlop/s	Japan
No. 2	Summit	Cores: 2414592 Rmax: 148600 TFlop/s	United States
No. 3	Sierra	Cores: 1572480 Rmax: 94640 TFlop/s	United States
No. 4	Sunway TaihuLight	Cores: 10649600 Rmax: 93015 TFlop/s	China
No. 5	Perlmutter	Cores: 761856 Rmax: 70870 TFlop/s	United States
No. 6	Selene	Cores: 555520 Rmax: 63460 TFlop/s	United States
No. 7	Tianhe-2	Cores: 4981760 Rmax: 61445 TFlop/s	China
No. 8	JUWELS Booster Module	Cores: 449280 Rmax: 44120 TFlop/s	Germany
No. 9	HPC5	Cores: 669760 Rmax: 35450 TFlop/s	Italy
No. 10	Voyager-EUS2	Cores: 448448 Rmax: 30050 TFlop/s	United States

With the continuous development of supercomputers, remote sensing, big data, and other technologies, supercomputers are applied in more aspects of remote sensing big data processing. So far, compared with cloud computing and GPU research, the application of supercomputers in the field of remote sensing is relatively small, mainly because high-performance computers are characterized by high construction cost, high maintenance cost, high energy consumption, and also have poor ease of use and high threshold. At present, supercomputers are mostly used in high-tech fields and cutting-edge technology research, such as military, medical, energy, aerospace, astronomy, and physics. With the increase in the number of supercomputers and the reduction of the construction threshold, the application of supercomputers in the field of remote sensing big data needs to be further explored. Table 6.5 lists the applications of supercomputers for Remote Sensing Big Data processing in different directions.

Table 6.5. Some relevant studies in the 2000s.

Scholars/Time	Supercomputers	Research
Plaza *et al.* (2006)	Thunderhead	Spectral end elements are automatically extracted and a speedup of 99–234 is obtained when using 256 processors
Osterloh *et al.* (2009)	MareNostrum	AOD inversion based on LIDAR data
Petcu *et al.* (2011)	IBM BlueGene/P	Fuzzy C-mean clustering algorithm
Kumar *et al.* (2011)	Cray XT5	Unusual Forest Events Mountain Pine Beetle Outbreak in Colorado, USA
Wu *et al.* (2013)	Tianhe 1A	A physics-based ray-tracing approach to simulate satellite images
Cai *et al.* (2018)	ROGER and Blue Waters	Crop type classification based on time-series Landsat data and deep learning methods
Kurte *et al.* (2019)	Titan	Human settlements covering an area of 685675 km^2 were mapped using 0.31 m resolution Worldview3 remote sensing imagery, for a dataset of size 13 TB, with a processing time of only 6 hours

6.4 GPU

Graphics Processing Unit (GPU), or graphics processor, has a higher parallel architecture and greater advantages in image and graphics processing compared to general-purpose processor CPUs.

GPUs date back to 1981 when IBM introduced the CGA, or Computer Graphics Processing Unit. 1994 saw the introduction of Matrox Impression, the first 3D graphics gas pedal for PCs. 1999 saw the release of Nvidia's Geforce256, which Nvidia itself claimed was the world's first GPU. In 2006, Microsoft and Nvidia introduced DirectX 10 and CUDA, respectively, enabling users to use GPU computing power for general-purpose computing (GPGPU). In 2009, OpenCL 1.0 was released; OpenGL is a framework for writing programs for heterogeneous platforms that can be composed of CPUs, GPUs, or other types of processors and is widely available for almost all graphics cards. AMP, Microsoft's new extension to the C++ programming language, accelerates the execution of C++ code by leveraging data-parallel hardware such as graphics processing units (GPUs). The key time points in the development of GPU technology are shown in Figure 6.3.

GPU's powerful low floating point power consumption, strong computing power, small size, and light gravity make it a cost-effective solution for remote sensing data processing. The high-dimensional and spatial geometric characteristics of remote sensing data are perfectly matched with GPU computing mode, and the use of GPU for remote sensing data processing can give full play to its highly parallel computing capability, where Control is the controller, ALU is the arithmetic logic unit, Cache is

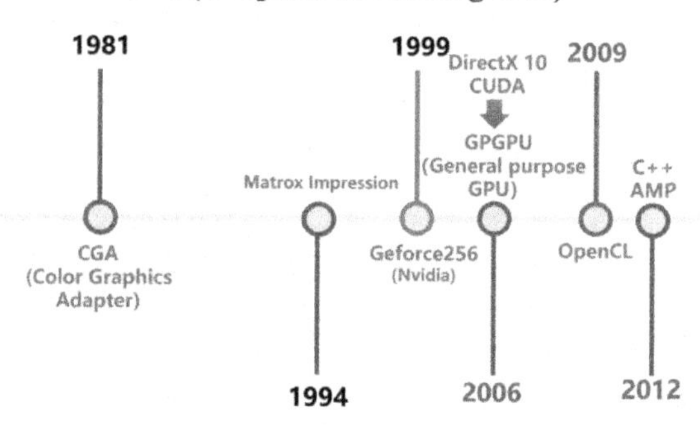

Figure 6.3. Development of GPU.

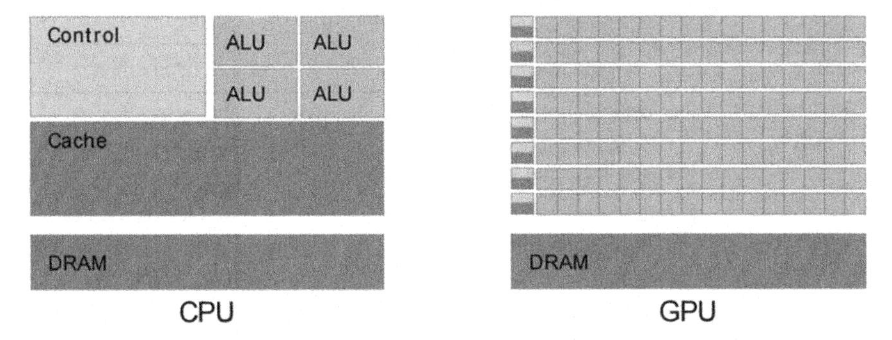

Figure 6.4. Comparison of CPU and GPU hardware logic structure.

the CPU internal cache, and DRAM is the memory. Figure 6.4 shows a schematic comparison of the CPU and GPU hardware logic structures. We can see that GPU designers use more transistors as execution units rather than as complex control units and caches as CPUs do. In practical terms, 5% of the CPU chip space is ALU, while 40% of the GPU space is ALU. This is what leads to the superior computational power of the GPU.

With the release of the first version of CUDA in 2007 and OpenCL in 2009, the programming model of GPU has been greatly simplified and GPU has been widely used in many aspects of remote sensing data processing. GPU computing can be applied in many aspects of remote sensing data image classification, decoding, change detection, segmentation, etc., and very significant computational efficiency improvement has been obtained as shown in Table 6.6.

At the same time, NVIDIA provides CUDA-X libraries for mathematics, artificial intelligence, and high-performance computing for GPU acceleration when developing remote sensing data in parallel with GPUs. The support of these professional libraries greatly reduces the difficulty of coding for developers, and CUDA's good ecosystem and ease of use support it to be the first choice for researchers using GPUs for remote sensing data processing.

OpenCL is a framework for writing programs for heterogeneous platforms, which can support multiple types of CPUs, GPUs, or other types of processors. Due to its better versatility, many researchers choose OpenCL for parallel development of GPUs. At this stage, CUDA and OpenCL are the two most widely used tools for GPU development. Dori (2021) compared the technical features of CUDA and OpenCL, as shown in Table 6.7. C++AMP and OpenACC are also capable of GPU

Table 6.6. Efficiency benefits.

Scholars	Specific methods	Accelerated effect
Češnovar *et al.* (2013)	Fast orthophoto generation using GPU clusters, open-source tool Orfeo Tool Box (OTB)	Relative single-threaded computing speedup of 198x
Zhen *et al.* (2014)	Parallel computation of semantic classification of remote sensing images	Tesla C1060 GPU is used to obtain a 39x speedup relative to the Core i5 650 CPU
López-Fandiño *et al.* (2017)	Extraction method of multitemporal hyperspectral remote sensing data based on watershed segmentation technique	Relative OpenMP speedup of 46.5x
Quesada-Barriuso *et al.*, (2021)	Parallel computation of hyperpixel segmentation for hyperspectral images	Relative to 8-threaded OpenMP, the GPU achieves a speedup ratio between 12.1x and 32.9x for different numbers of partitions

Table 6.7. CUDA vs. OpenCL.

Comparison	CUDA	OpenCL
Performance	No clear advantage, dependent code quality, hardware type, and other variables	No clear advantage, dependent code quality, hardware type, and other variables
Vendor Implementation	Implemented by only NVIDIA	Implemented by TONS of vendors including AMD, NVIDIA, Intel, Apple, Radeon, etc.
Portability	Only works using NVIDIA hardware	Can be ported to various other hardware as long as vendor-specific extensions are avoided
Open-source vs. Commercial	Proprietary framework of NVIDIA	Open-source standard
OS Support	Supported on the leading operating systems with the only distinction of NVIDIA hardware must be used	Supported on various Operating Systems
Libraries	Has extensive high-performance libraries	Has a good number of libraries which can be used on all OpenCL compliant hardware but not as extensive as CUDA

Table 6.7. (*Continued*)

Comparison	CUDA	OpenCL
Community	Has a larger community	Has a growing community not as large as CUDA
Technicalities	Not a language but a platform and programming model that achieves parallelization using CUDA keywords	Does not enable writing code in C++ but works in a C programming language resembling environment

Source: Dori (2021).

development, and since there are fewer studies related to remote sensing data processing based on other tools, this paper will not be expanded.

Deep collaborative computing of GPU and CPU has great potential in complex and heterogeneous big data processing and analysis. When CPU/GPU collaborative computing of remote sensing data processing problems with multitask processes, data dependency between sub-tasks increases the cost of data exchange between GPU and CPU. So, it is important to use CPU and GPU resources rationally in processing and analysis of Remote Sensing Big Data.

Although researchers have obtained high acceleration ratios by using GPU acceleration, the acceleration efficiency varies very much. To fully utilize the computational performance of GPUs, many researchers have optimized the parallel computing strategy of GPUs in terms of memory management, data exchange, task scheduling, and combination with other parallel computing techniques, as shown in Table 6.8.

The natural advantages of GPU in parallel processing of remote sensing data have made it widely used in many aspects such as fusion, resampling, image classification, image compression, image retrieval, cloud processing, target detection, and inversion of remote sensing big data, and the application scope and application direction are growing year by year. However, the parallel strategies and algorithms of GPUs are highly coupled and difficult to reuse, which leads to relatively high development costs. Therefore, some researchers have conducted some explorations on the generality of GPU development, for example, Ma *et al.* proposed a parallel processing model for remote sensing images with reusable GPUs by designing GPU parallel programming templates and intermediate transition data models, which provides developers with a simple effective parallel programming method for remote sensing image processing (Ma *et al.*, 2014a).

Table 6.8. Parallel computing strategy optimization for GPU.

Research Scholars	Optimization Methods
Liu *et al.* (2014)	Data buffer, texture memory, multithreading, and other technologies are optimized for data loading, memory usage, and data transfer, solving the problems of a large number of loading limits, low bandwidth of GPU memory, and host memory data transfer.
Sun *et al.* (2014)	Optimization strategies to maximize device occupancy, improve memory access efficiency, and increase instruction throughput are proposed for remote sensing image correction based on improved rational function models when implemented using CPU/GPU synergy.
Zhang and Li (2015)	Dynamic task scheduler based on reinforcement learning for dynamic and automatic optimal allocation of computational resources for LiDAR data processing.
Liu *et al.* (2019)	A three-level parallelism strategy of job-level parallelism, task-level parallelism, and image-processing-level parallelism is proposed. The first two levels are coarse-grained parallelism, which is realized by multi-DAG, and the image-processing level is fine-grained parallelism for long time series Aerosol Optical Depth (AOD) inversion.
Ordonez *et al.* (2017)	Improvement strategies are proposed for reducing host and device data transfers, reducing new memory allocations, using more texture memory, using more CUDA libraries, using more low-level instructions, and optimizing core configurations for hyperspectral image alignment.

With the increasing computational volume of Remote Sensing Big Data, energy consumption gradually becomes an issue that cannot be ignored, and GPU has higher energy efficiency under the same computational power measurement. The lower-energy consumption and powerful processing capability make GPU the preferred choice for scenarios such as onboard real-time processing. Under the same budgetary conditions of hardware cost, multicore CPU parallel algorithms based on MPI, OpenMP, etc. cannot match GPU parallel algorithms in most cases due to the obvious advantage in the number of GPU cores. With the continuous development of GPU development technology and the continuous improvement of program ecosystem, the application of GPU in the field of Remote Sensing Big Data processing will become more and more widespread.

6.5 OpenMP

OpenMP is a programming model on the shared storage architecture, which has the advantages of simplicity, portability, and scalability. It provides abstract descriptions of the high-level parallel algorithm, which can easily realize the parallelization and multithreading of the program, thereby reducing the difficulty and complexity of parallel programming, as shown in Figure 6.5.

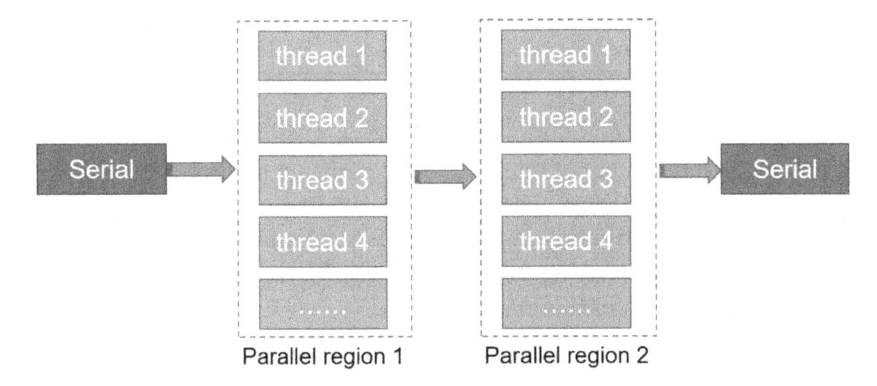

Figure 6.5. OpenMP parallel computing schematic.

With the popularization of multicore CPU, some researchers (as shown in Table 6.9) choose OpenMP to achieve parallel processing when performing high-performance computing of remote sensing data.

Table 6.9. Application of OpenMP in remote sensing data processing.

Scholars	Content	Efficiency
Hu *et al.* (2015)	Parallel Computation Optimization of SAR Data Echo Simulation Based on OpenMP and Vector Instruction Set	Compared with about 23 times higher than the traditional method
Zheng *et al.* (2016)	Multicore processor implementation of inverse distance weighted interpolation and bilinear interpolation algorithm in aerosol optical thickness inversion based on OpenMP	For images of different sizes, the speedup of 8-thread parallel method is between 3.24–3.61 and 3.72–6.69
Zhu *et al.* (2016)	OpenMP is used to implement the FLICM algorithm for SAR unsupervised remote sensing image change detection on Intel MIC (multi-integrated kernel)	A speedup of 20× over single CPU serial computation

(*Continued*)

Table 6.9. (*Continued*)

Scholars	Content	Efficiency
Mullah and Deka (2018)	A multicore parallel algorithm for super-resolution reconstruction of remote sensing images based on sparse matrix is developed using PARAMTEZ supercomputer and OpenMP	A speedup of 10.83 relative to a single core using 24 cores
Du *et al.* 2017)	Large amount of super high-resolution satellite image matching	A speedup of 12.3 times over a single thread with 16 threads

Compared with other tools, OpenMP has a relatively limited acceleration effect on the process of massive remote sensing images due to its own technical characteristics. Therefore, there are few studies on the use of OpenMP parallel computing in the field of large remote sensing data processes. Mixed programming with MPI and GPU is often used to achieve higher computational efficiency.

6.6 Message Passing Calculation

Many application programming interfaces and implementation libraries supporting computer serial process communication models are called message passing models. So far, Message Passing Interface (MPI) has been one of the most important parallel programming tools based on message passing, which has the advantages of good portability, powerful function, and high efficiency and has a variety of different free, efficient, and practical implementation versions. Almost all parallel computer manufacturers provide support for it, which has become a *de facto* message passing parallel programming standard (Chen, 2011).

MPI has been continuously updated and expanded since its appearance in the 1990s. With hundreds of commands, MPI can cope with most situations in parallel computing. Now, MPI is still the main model of high-performance computing. The two core issues of parallel computing programming using MPI are task execution and data exchange. Reasonable task allocation, task scheduling, and communication schemes can improve the performance of parallel computing. MPI standard defines the syntax and semantics of the core library, which can be called by Fortran and C to form a portable information transfer program. MPI provides the basis set for adapting to various parallel hardware vendors, which are effectively implemented. This leads to hardware vendors who can create high-level practices based on this series of underlying standards, thus providing their parallel machines for distributed memory interaction systems. MPI provides a simple and easy-to-use portable interface that is powerful enough for programmers to perform high-performance information transfer operations on advanced machines.

Considering the characteristics of remote sensing data processing and analysis, such as dense data and complex calculation steps, scholars have also formed some achievements in data transmission, I/O operation, debt balance, architecture optimization, and system versatility. There is a big gap between the performance of a processor and an I/O system. Overlapping I/O operation and calculation to reduce I/O time consumption and frequent data communication between nodes can improve the efficiency of parallel computing.

When heterogeneous cluster is used, computing resource load balancing is an important factor affecting the computational efficiency, so it is also a factor to be considered in task allocation. The classical framework of multinode or multicore CPU parallel processing system in computer cluster environment is the master/slave architecture. According to the different roles in the cluster, the computer is divided into master node and slave node. The master node is responsible for task division, task allocation, task merging, and other management. The slave node is responsible for calculation. The typical three-step operation strategy of the master/slave architecture is as follows: The first step is that the master node is responsible for dividing tasks and sending tasks to the slave node. The second step is that the task is received from the node and calculated, and the results are sent to the main node. The third step is main node merging and output results (Gu *et al.*, 2018; Wang *et al.*, 2011). Researchers have optimized MPI from different aspects of MPI parallel computing, and the related research strategies are shown in Table 6.10.

Most parallel processing frameworks or systems have poor versatility and poor portability. To improve the versatility of parallel systems, some scholars have designed a parallel computing framework with relatively strong versatility from the perspectives of parallel architecture, data model, and processing function expansion. Using the general strong parallel computing framework, developers do not need to invest too much energy in the implementation of parallel processing and can quickly achieve efficient parallel implementation of different remote sensing algorithms.

Table 6.10. Optimization measures.

Optimizing direction	Specific method	Scholars
Data storage	Using a disk sharing cluster, the results of each child task are written to the same pyramid file	He *et al.* (2015)
Data transmission	Connecting 30 Parallel Computing Nodes with Transmission Speed of 40 Gb/s	Du *et al.* (2016)
Storage and transmission	The strategy of relative separation of communication and processing is used, and the data storage strategy and network transmission bandwidth are optimized	Tie *et al.* (2018)
Parallel strategy	A dual parallel strategy of data and task is adopted to reduce the overall running time by overlapping computation and data transmission	Pu *et al.* (2019)
Load balancing	Load balancing of heterogeneous computing resources in computer cluster using workload estimation algorithm	Plaza *et al.* (2010)
Parallel strategy	DAG scheduling is used to realize inter-task parallelism and MPI is used to realize intra-task parallelism	Ma *et al.* (2014)

6.7 FPGA

Field Programmable Gate Array (FPGA) is a reconfigurable circuit chip and a hardware reconfigurable architecture. Through programming, users can change its application scenarios at any time; it can simulate CPU, GPU, and other hardware parallel computing. By interconnecting with the high-speed interface of the target hardware, FPGA can complete the low-efficiency part of the target hardware, thus accelerating at the system level. The logic architecture and the integrated circuit board of the FPGA are shown in Figure 6.6.

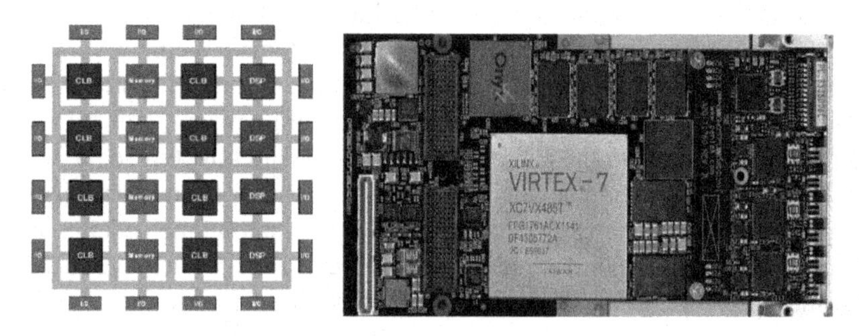

Figure 6.6. FPGA architecture schematic and integrated circuit board.

FPGA has the advantages of small size, light-weight, and low energy consumption, which is especially suitable for the application in the scenario of limited energy consumption and load budget of platforms such as satellites or aircraft. In addition, compared with CPU and GPU computing, FPGA has obvious energy efficiency advantages while obtaining high-performance computing efficiency, which can be used as a choice for high-performance computing in low energy consumption scenarios. Some researchers have summarized the technical advantages of FPGA in remote sensing big data processing applications, some related literature, and as shown in Table 6.11.

Table 6.11. FPGA application in remote sensing big data processing.

Technological superiority	Content	Scholar
Spaceborne real-time processing	The cloud detection algorithm is implemented based on HPC-I platform. HPC-I is designed based on Xilinx XQR 4062 FPGA. The payload is small, the size is 195x165x35mm, the power consumption is about 2.5W, and the speed processing is 1.3ms (256 × 256I 50Mpix/s), which effectively reduces the space storage and downlink capacity requirements.	Williams *et al.* (2002)
	A remote sensing image compression algorithm based on adaptive JPEG-LS is implemented on the same HPC-I platform to meet the needs of real-time spaceborne image compression.	Visser *et al.* (2003)
	Based on FPGA and YOLOv3 — tiny target detection acceleration algorithm — a high-performance and low-power hardware accelerator for remote sensing image onboard real-time target detection is realized by parallelizing each convolution layer to improve resource utilization and calculation speed.	Zheng *et al.* (2020)
Low-energy consumption	A high-performance deep convolutional neural network (DCNN) remote sensing image classification accelerator based on FPGA is proposed. By using dual-channel dual-data rate synchronous dynamic random access memory (DDR) access mode and on-chip data processing optimization strategy, an efficient hardware architecture is obtained. Compared with CPU and	Zhang *et al.* (2020)

(*Continued*)

Table 6.11. (*Continued*)

Technological superiority	Content	Scholar
	GPU, the energy efficiency of the accelerator is increased by 174.53 times and 11.51 times, respectively.	
	A hybrid type reasoning method based on symmetric quantization scheme is proposed for remote sensing image convolutional neural network (CNN) classification based on FPGA. By using low bit width integers instead of floating points, the power consumption of lookup table (LUT), trigger (FF), digital signal processor (DSP), and block random access memory (BRAM) is reduced by 46.21%, 43.84%, 45%, and 51%, respectively, compared with floating point implementation.	Wei *et al.* (2019)
	An improved YOLO v2 remote sensing image target detection method is implemented on FPGA, and a hardware/software collaborative framework with good versatility and flexibility is designed. Compared with CPU, the energy consumption is only 5.96 W while the speedup is 33.4x.	Zhang *et al.* (2021)

6.8 Remote Sensing Grid Computing

As a distributed computing infrastructure to meet advanced science and engineering applications, "Grid" was proposed in the mid-1990s. Due to the high cost of supercomputers and the need for professional maintenance, the development of grid computing has overcome these shortcomings. It is of great significance to study the rapid processing of massive remote sensing data by using grid computing (Xue *et al.*, 2007).

After the introduction of grid computing technology, many countries or regions have launched their own grid projects. Asia-Pacific Grid, ApGrid, is an Asia-Pacific Partnership Grid Computing project organized by an open community collaboration that supports large-scale applications of Earth science and biological information; NASA's Information Power Grid (IPG) grid project is a high-performance computing and data grid project for NASA scientists and engineers. It relies on large-scale decentralized resources to provide users with a sustainable, large-scale, multi-agency/multidisciplinary problem-solving environment. The Enabling Grids for E-SciencE (EGEE) is the follow-up of DATAGrid project. Many technologies and infrastructures of DATAGrid are included in the EGEE project, aiming to develop grid infrastructure across Europe in the fields of high energy physics, earth science, and life science. The TeraGrid grid project in the United States is a national grid project supported by the National Science Foundation of the United States (NSF). Its design goal is to promote the encapsulation of data capabilities, computing capabilities, and network capabilities and provide a platform for grid communities to carry out data-intensive applications and more integrated computing-intensive applications. The Grid Enabled Integrated Earth System Model (GENIE) grid integrated earth system model project in the UK is supported by the UK Natural Environment Research Council through the e-Science project. Its purpose is to establish an easy-to-use and fast-running earth climate system model based on the grid computing framework so that users can share the distributed data in the grid environment, run the model, and obtain the analysis results. The Earth System Grid II (ESG II) project of the United States is a research project of the United States Department of Energy. Its main goal is to provide a seamless and powerful environment using grid technology and other emerging technologies to integrate distributed supercomputers, large-scale data and analysis servers, to solve the severe challenges related to the development of global geographic system model analysis and knowledge, and to make the next generation of climate research possible. The Grid Processing on Demand (G-POD) of the European Space Agency (ESA) is a dedicated ground observation grid infrastructure project hosted by the European Space Agency (ESA). By coupling with the dedicated high-performance cluster, it provides a flexible way to build a virtual application environment so that users can seamlessly "insert" the dedicated data processing applications and quickly obtain data, computing resources, and results.

As shown in Table 6.12, researchers and engineers have carried out some Remote Sensing Big Data processing work based on Grid System. Remote Sensing Big Data processing based on grid computing can integrate heterogeneous computing resources well. By optimizing and sharing these resources, it has obvious cost advantages over high-performance computing platforms such as supercomputers. However, the application of grid computing in remote sensing big data needs to solve the key

Table 6.12. Grid computing system for high-performance computing of remote sensing data.

Grid system	Remote sensing research	Research scholar
DataGrid	Global ozone monitoring	Nico *et al.* (2003)
China Grid	Geometric correction and image registration of remote sensing data	Zhou *et al.* (2006/2007)
Medio Grid	Prevention of floods and forest fires	Petcu *et al.* (2007)
WIDE AREAGRID	Flood Analysis Using Envisat and Radarsat-1 Images	Kussul *et al.* (2012) Grid System

technologies such as computing mode, resource management, task scheduling, data organization, and stability. There are still the following problems: not all remote sensing processing algorithms are suitable for grid parallelization; for the improvement of some processing algorithms, it takes a huge amount of work to achieve the purpose of large throughput; grid configuration affects the performance, reliability, and security of computing infrastructure in large programs (Xue *et al.*, 2008; Zhu *et al.*, 2016).

6.9 Big Data and Cloud Computing

Traditional parallel applications based on computer clusters mostly use MPI to achieve parallelism, but programming with MPI is more complicated for remote sensing researchers without computer basics. With the emergence and development of cloud computing technology, large-scale distributed computing for remote sensing big data has formed a more mature ecology, reducing the difficulty of building high-performance computing platforms. In addition, the distributed file system is one of the advantages of cloud computing in remote sensing big data processing, as shown in Figures 6.7 and 6.8. And more and more researchers are using cloud computing technology for remote sensing big data processing and analysis. Hadoop and Spark are typical platforms that are widely used, and many scholars have conducted many studies based on these two platforms in improving the computational efficiency of remote sensing big data processing and analysis tasks.

Hadoop originated as a search application developed by Google and later developed as an open-source system by the Apache Foundation.

Figure 6.7. Benefits of big data cloud computing technology.

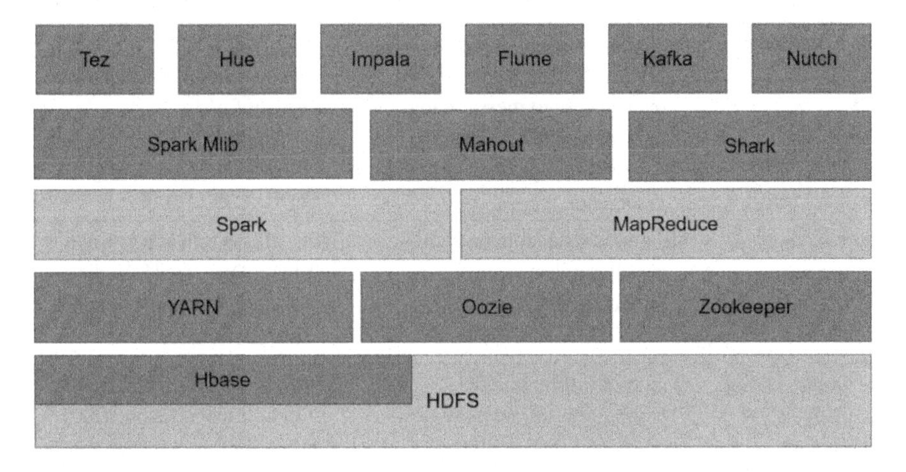

Figure 6.8. Ecosystem of Hadoop and Spark.

The core of Hadoop is Hadoop Distributed File System (HDFS) and MapReduce (Mapper + Reducer), where HDFS is used for massive data storage and MapReduce is used for massive data computation. The Hadoop parallel computing framework is suitable for a variety of scientific applications geared towards remote sensing data processing services.

Although Hadoop has been widely used for parallel processing of large-scale remote sensing data since its launch, there is an inability to solve the problems of intensive disk I/O operations and frequent network communication, which makes it difficult to further meet the needs of application scenarios such as real-time processing and analysis of massive and heterogeneous remote sensing data (Huang *et al.*, 2017). Apache

Spark is a fast-general-purpose computing engine designed for large-scale data processing. Hadoop's MapReduce writes data back to physical storage media after each operation, while Spark performs most operations in memory, giving Spark a significant speed advantage over Hadoop.

Big Data technology involves Big Data access, Big Data preprocessing, Big data computing, Big Data storage, data analysis, and mining, Big Data sharing and exchange, Big Data presentation, and many other aspects, as shown in Table 6.13.

Table 6.13. Big data technologies and related tools.

	Technical content	Related technologies (tools)
Big Data access	Access to existing data, access to real-time data, access to file data, access to message log data, access to text data, access to image data, access to video data	Kafka, ActiveMQ, ZeroMQ, Flume, Sqoop, Socket(Mina, Netty), ftp/sftp
Big Data storage	Structured data storage, semi-structured data storage, unstructured data storage	Hdfs, Hbase, Hive, S3, Kudu, MongoDB, Neo4J, Redis, Alluxio (Tachyon), Lucene, Solr, ElasticSearch
Data processing analysis mining	Data processing, parallel computing, offline analytics, quasi-real-time analytics, real-time analytics, image recognition, speech recognition, machine learning	MapReduce, Hive, Pig, Spark, Flink, Impala, Kylin, Tez, Akka, Storm, S4, Mahout, MLlib, CUDA, OpenCL
Big Data sharing and exchange	Data access, data cleansing, conversion, de-sensitization, decryption, data asset management, data export	Kafka, ActiveMQ, ZeroMQ, Dubbo, Socket(Mina, Netty), ftp/sftp, RestFul, Web Service
Big Data presentation	Graphical presentation (scatter, line, bar, map, pie, radar, K-line, box line, heat map, relationship, rectangular tree, parallel coordinates, Sankey, funnel, dashboard), text presentation	Echarts, Tableau

When using big data processing computing engines, in order to meet the needs of high-performance computing for remote sensing big data processing, some researchers have carried out special research on efficient storage, organization, management, and distribution of remote sensing data based on distributed file systems. The widespread use of big data has driven the rapid update of cloud computing technology, with new technologies being introduced and different technical architectures being applied to different application scenarios, some relevant research works are shown in Table 6.14. The cloud computing framework has a more complete ecosystem than MPI-based distributed parallel computing in a cluster environment. The cloud computing framework implements the task scheduling that is a key consideration in the MPI process through a dedicated task scheduling system.

Table 6.14. Remote sensing big data applications.

Research scholars	Research directions	Research methodology
Rathore *et al.* (2016)	Real-time feature extraction and detection of remote sensing satellite images	The Hadoop ecosystem is used to process the entire process of remote sensing data of chunking, computing, merging, and managing, achieving near real-time computing efficiency.
Zou *et al.* (2018)	Global vegetation drought status monitoring	An abstract data format achieves a unified description of remote sensing data by discretizing multidimensional remote sensing data for distributed storage and computation and obtains a near-linear reduction in time overhead and a near-linear increase in throughput by using MapReduce to solve the complexity of remote sensing algorithms.
Wu *et al.* (2016)	Hyperspectral dimensionality reduction principal component analysis	A parallel algorithm based on cloud computing architecture is designed, using Spark as the computational engine and HDFS for distributed storage, in order to fully exploit the high throughput access and high-performance distributed computing capabilities of the cloud computing environment.

In terms of efficiency, there is no significant difference between MPI and cloud computing; in terms of ease of use, the cloud computing framework performs better due to its relatively well-established ecosystem. In general, when using MPI to process remote sensing big data, it is necessary to consider the running state of processes, communication between different processes, and monitoring of anomalies, while the cloud computing framework for remote sensing big data has a seamless computing engine, a distributed file system, and a powerful task scheduling system. The framework provides a complete solution without the need for remote sensing researchers to spend too much effort on task management, load balancing, etc. Therefore, cloud computing framework is currently the preferred choice for remote sensing researchers to implement parallel computing based on computer clusters.

6.10 Remote Sensing Cloud Computing Platform

Both grid computing and cloud computing aim to integrate distributed and heterogeneous computing resources and provide high-performance computing capabilities. While grid computing systems tend to be application-specific or task-specific, cloud computing platforms are more general in nature, which is why some researchers refer to cloud computing as "grid computing 2.0". Cloud computing platform is an effective way to store, access, and analyzes datasets on very powerful servers that can virtualize supercomputers for users. To meet the research and application needs of the increasing volume of remote sensing data and the strong demand for massive remote sensing data processing capabilities, the development of remote sensing cloud computing technology and platforms have emerged to provide unprecedented opportunities for remote sensing big data processing and analysis: the cloud has massive data resources that do not need to be downloaded to local processing; the cloud provides batch and interactive big data computing services; the cloud provides application program interfaces, APIs, eliminating the need to install software locally (Fu *et al.*, 2020).

At present, the main remote sensing cloud computing platforms are Amazon Web Services (AWS), Azure, Google Earth Engine (GEE), Descartes labs, European Open Science Cloud (EOSC), Australian Geoscience Data Cube (AGDC), PIE-Engine, EarthDataMiner, Aliyun, etc. These cloud computing platforms integrate huge amounts of remote sensing data and can provide powerful arithmetic power as well as program interfaces for spatial data and remote sensing data processing and analysis, as shown in Table 6.15 (Lewis *et al.*, 2017; Dong *et al.*, 2020; Fu *et al.*, 2020; Amani *et al.*, 2020).

Table 6.15. Overview of representative remote sensing cloud computing platforms.

Remote sensing cloud computing platform	Integrated datasets	Development languages supported
Amazon Web Services (AWS)	Landsat-8, Sentinel-1, Sentinel-2, China-Pakistan Resource Satellite, etc.	JavaScript, Python, C++, etc.
Azure	Landsat, Sentinel-2, MODIS, etc.	C#, F#, Node.js, Python
Google Earth Engine (GEE)	Landsat, MODIS, Sentinel series, and several atmospheric, meteorological, and vector datasets.	JavaScript, Python
Descartes labs	Landsat, Sentinel, SPOT, and other remote sensing data as well as meteorological, elevation, and land use data.	Python
Australian Geoscience Data Cube (AGDC)	Landsat, MODIS, Sentinel, vegetation cover, elevation, land cover, etc.	Python
PIE-Engine	Landsat, Sentinel, Resource 3, High Score Series, Himawari-8 and population, meteorological data, etc.	JavaScript, Python
EarthDataMiner	MODIS, Landsat, Sentinel, ZY3, GF1/2, Bioecology, Atmospheric, etc.	Python

At present, Google Earth Engine (GEE) is one of the most widely used remote sensing cloud computing platforms. GEE, a cloud computing platform launched by Google in 2010, supports automatic parallel processing and fast computing and is currently the most popular platform for processing big data on Earth (Amani *et al.*, 2020). GEE integrates nearly 40 years of petabytes of historical remote sensing imagery and geospatial datasets, which is continually updated or expanded daily to provide global-scale analysis capabilities; scientists, researchers, and developers can use GEE for change detection, trend mapping, and quantifying differences in the Earth's surface; GEE is available for commercial use as well as free for academic and research use. GEE consists of datasets, computational resources, code editors, APIs, etc. GEE provides a simple and powerful programming interface library that supports popular coding languages such as JavaScript and Python, plus the ability for users to upload their own raster and vector data for processing and analysis.

The Earth Engine Code Editor at code.earthengine.google.com is a web-based IDE for the Earth Engine JavaScript API. It requires login with a Google Account that's been enabled for Earth Engine access. Code Editor features are designed to make developing complex geospatial workflows fast and easy. The Code Editor has the following elements (illustrated in Figure 6.9).

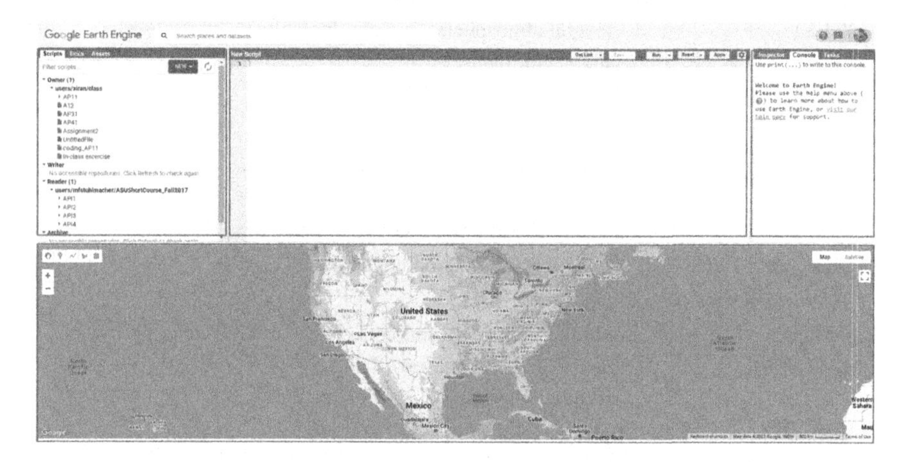

Figure 6.9. GEE code editor.

The Explorer is a simple web interface to the Earth Engine API, as shown in Figure 6.10. It allows anyone to visualize the data in the public data catalog. Signed in Earth Engine users can also import data, run simple analyses, save, and export the results.

China Aerospace Information Technology Company has developed a series of products for the PIE Engine (as shown in Figure 6.11), including the remote sensing computing cloud service platform PIE-Engine Studio, which is a geospatial data analysis and computing platform built on the top of cloud computing. PIE provides free, flexible, and elastic computing services for scientific researchers and engineers in large-scale geographic

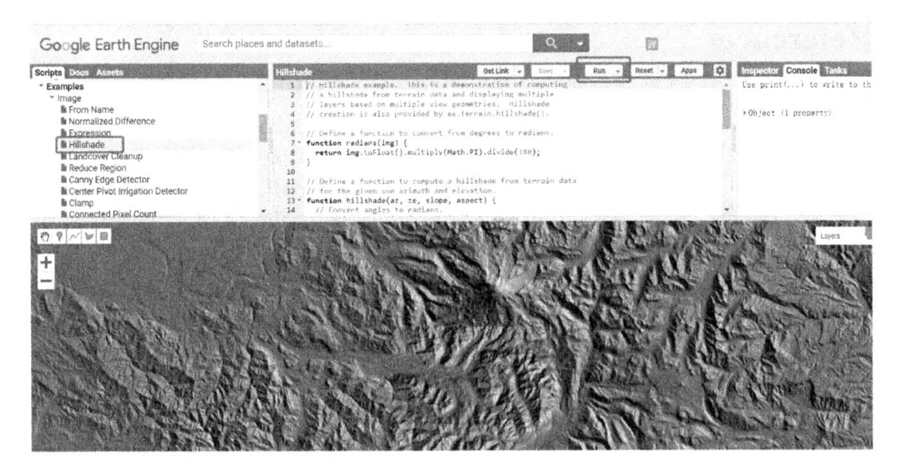

Figure 6.10. GEE explorer.

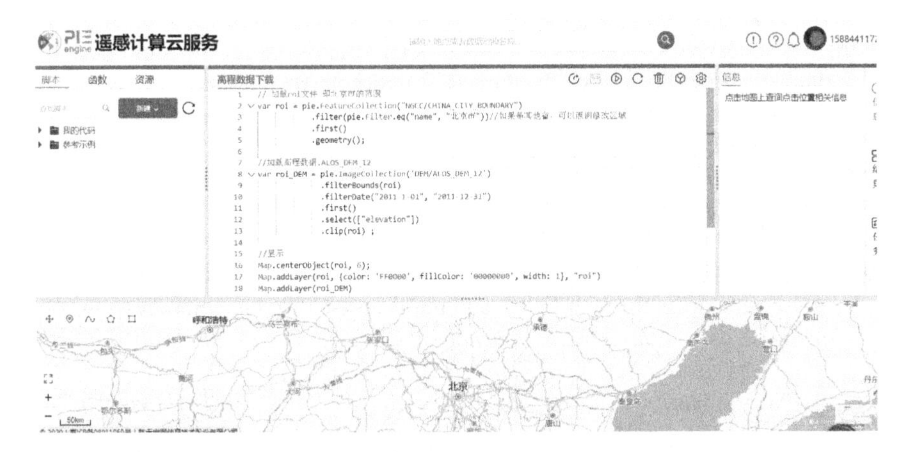

Figure 6.11. PIE remote sensing cloud computing service.

data analysis and scientific research. By combining spatial data such as massive satellite remote sensing images and geographic element data, PIE-Engine Studio enables users to study algorithmic models at any scale based on the platform and take interactive programming to validate them, enabling rapid exploration of surface features and discovery of changes and trends.

References

Amani, M., Ghorbanian, A., Ahmadi, S. A., Kakooei, M., Moghimi, A., Mirmazloumi, S. M., Moghaddam, S. H. A., Mahdavi, S., Ghahremanloo, M., Parsian, S., Wu, Q., and Brisco, B. (2020). Google Earth Engine cloud computing platform for remote sensing big data applications: A comprehensive review. *IEEE Journal of Selected Topics in Applied Earth Observations and Remote Sensing*, **13**: 5326–5350.

Armstrong, M. P. and Marciano, R. (1994). Inverse-distance-weighted spatial interpolation using parallel supercomputers. *Photogrammetric Engineering and Remote Sensing*, **60**: 1097–1102.

Aspri, M., Tsagkatakis, G., and Tsakalides, P. (2020). Distributed training and inference of deep learning models for multi-modal land cover classification. *Remote Sensing*, **12**(17): 2670. https://doi.org/10.3390/rs12172670.

Atoche, A. C., Castillo, J. V., Aguilar, J. O., Álvarez, R. C., and Viñas, J. A. (2020). Intelligent classification of large-scale remotely sensed hyperspectral images using multi-GPU computing. *IEEE Latin America Transactions*, **18**: 113–119.

Bernabe, S., Garcia, C., Igual, F. D., Botella, G., Prieto-Matias, M., and Plaza, A. (2019). Portability study of an OpenCL algorithm for automatic target detection in hyperspectral images. *IEEE Transactions on Geoscience and Remote Sensing*, **57**: 9499–9511.

Bernabe, S., Jimenez, L. I., Garcia, C., Plaza, J., and Plaza, A. (2018). Multicore real-time implementation of a full hyperspectral unmixing chain. *IEEE Geoscience and Remote Sensing Letters*, **15**: 744–748.

Bilotta, G., Sanchez, R. Z., and Ganci, G. (2013). Optimizing satellite monitoring of volcanic areas through GPUs and multi-core CPUs image processing: An OpenCL case study. *IEEE Journal of Selected Topics in Applied Earth Observations and Remote Sensing*, **6**: 2445–2452.

Boulila, W., Sellami, M., Driss, M., Al-Sarem, M., Safaei, M., and Ghaleb, F. A. (2021). RS-DCNN: A novel distributed convolutional-neural-networks based-approach for big remote-sensing image classification. *Computers and Electronics*, **182**: 106014. https://doi.org/10.1016/j.compag.2021.106014.

Cai, X. W., Hu, J., Li, Z. Y., and Zhu, B. (2012). A universal parallel framework for remote sensing process. *Applied Mechanics and Materials*, **239–240**: 599–602.

Cai, Y., Guan, K., Peng, J., Wang, S., Seifert, C., Wardlow, B., and Li, Z. (2018). A high-performance and in-season classification system of field-level crop types using time-series Landsat data and a machine learning approach. *Remote Sensing of Environment*, **210**: 35–47.

Češnovar, R., Risojević, V., Babić, Z., Dobravec, T., and Bulić, P. (2013). A GPU implementation of a structural-similarity-based aerial-image classification. *The Journal of Supercomputing*, **65**: 978–996.

Chang, Y.-L., Chen, Z.-M., Liu, J.-N., Chang, L., and Fang, J. P. (2010). Parallel K-dimensional tree classification based on semi-matroid structure for remote sensing applications. In *Satellite Data Compression, Communications, and Processing VI*, p. 78100S. International Society for Optics and Photonics.

Chen, F., Wang, N., Yu, B., Qin, Y., and Wang, L. (2021). A strategy of parallel seed-based image segmentation algorithms for handling massive image tiles over the spark platform. *Remote Sensing*, **13**(10): 1969. https://doi.org/10.3390/rs13101969.

Chi, M., Plaza, A., Benediktsson, J. A., Sun, Z., Shen, J., and Zhu, Y. (2016). Big data for remote sensing: Challenges and opportunities. *Proceedings of the IEEE*, **104**: 2207–2219.

Dong, H., Li, T., Leng, J., Kong, L., and Bai, G. (2017). GCN: GPU-based cube CNN framework for hyperspectral image classification. In *2017 46th International Conference on Parallel Processing (ICPP)*.

Du, S., Wang, M., and Fang, S. (2017). Block-and-octave constraint SIFT with multi-thread processing for VHR satellite image matching. *Remote Sensing Letters*, **8**: 1180–1189.

Du, Z., Gu, Y., Zhang, C., Zhang, F., Liu, R., Sequeira, J., and Li, W. (2016). ParSymG: A parallel clustering approach for unsupervised classification of remotely sensed imagery. *International Journal of Digital Earth*, **10**: 471–489.

Fang, L., Wang, M., Li, D., and Pan, J. (2014). CPU/GPU near real-time preprocessing for ZY-3 satellite images: Relative radiometric correction, MTF compensation, and geocorrection. *ISPRS Journal of Photogrammetry and Remote Sensing*, **87**: 229–240.

Garea, A. S., Heras, D. B., and Argüello, F. (2018). Caffe CNN-based classification of hyperspectral images on GPU. *The Journal of Supercomputing*, **75**: 1065–1077.

Giordano, R. and Guccione, P. (2017). ROI-based on-board compression for hyperspectral remote sensing images on GPU. *Sensors* (Basel), **17**(5): 1160. doi: 10.3390/s17051160.

Golpayegani, N. and Halem, M. (2009). Cloud computing for satellite data processing on high end compute clusters. In *2009 IEEE International Conference on Cloud Computing*.

González, C., Resano, J., Mozos, D., Plaza, A., and Valencia, D. (2010). FPGA implementation of the pixel purity index algorithm for remotely sensed hyperspectral image analysis. *EURASIP Journal on Advances in Signal Processing*, **2010**: 969806. https://doi.org/10.1155/2010/969806.

González, C., Sánchez, S., Paz, A., Resano, J., Mozos, D., and Plaza, A. (2013). Use of FPGA or GPU-based architectures for remotely sensed hyperspectral image processing. *Integration*, **46**: 89–103.

Gu, H., Han, Y., Yang, Y., Li, H., Liu, Z., Soergel, U., Blaschke, T., and Cui, S. (2018). An efficient parallel multi-scale segmentation method for remote sensing imagery. *Remote Sensing*, **10**(4): 590; https://doi.org/10.3390/rs10040590.

He, G., Wei, X., Chen, L., Wu, Q., and Jing, N. (2015). A MPI-based parallel pyramid building algorithm for large-scale remote sensing images. In *23rd International Conference on Geoinformatics*, pp. 1–4. IEEE.

Hossam, M. A., Ebied, H. M., Abdel-Aziz, M. H., and Tolba, M. F. (2014). Accelerated hyperspectral image recursive hierarchical segmentation using GPUs, multicore CPUs, and hybrid CPU/GPU cluster. *Journal of Real-Time Image Processing*, **14**: 413–432.

Hu, C., Zhang, F., Ma, L., Li, G., Hu, W., and Li, W. Efficient SAR raw data parallel simulation based on multicore vector extension. In *2015 IEEE International Geoscience and Remote Sensing Symposium (IGARSS)*, pp. 4719–4722. IEEE.

Huang, W., Meng, L., Zhang, D., and Zhang, W. (2017). In-memory parallel processing of massive remotely sensed data using an apache spark on hadoop YARN model. *IEEE Journal of Selected Topics in Applied Earth Observations and Remote Sensing*, **10**: 3–19.

Jing, W., Huo, S., Miao, Q., and Chen, X. (2017). A model of parallel mosaicking for massive remote sensing images based on spark. *IEEE Access*, **5**: 18229–18237.

Kumar, J., Mills, R. T., Hoffman, F. M., and Hargrove, W. W. (2011). Parallel k-means clustering for quantitative ecoregion delineation using large data sets. *Procedia Computer Science*, **4**: 1602–1611.

Kurte, K., Sanyal, J., Berres, A., Lunga, D., Coletti, M., Yang, H. L., Graves, D., Liebersohn, B., and Rose, A. (2019). Performance analysis and optimization for scalable deployment of deep learning models for country-scale settlement mapping on Titan supercomputer. *Concurrency and Computation: Practice and Experience*, **31**. https://doi.org/10.1002/cpe.5305.

Kussul, N. N., Shelestov, A. Y., Skakun, S. V., Li, G., and Kussul, O. M. (2012). The wide area grid testbed for flood monitoring using earth observation data. *IEEE Journal of Selected Topics in Applied Earth Observations and Remote Sensing*, **5**: 1746–1751.

Lee, K.-S. and Kang, H.-S. (1990). Assessing sea surface temperature in the yellow Sea using satellite remote sensing data. *Korean Journal of Remote Sensing*, **6**: 39–47.

Lewis, A., Oliver, S., Lymburner, L., Evans, B., Wyborn, L., Mueller, N., Raevksi, G., Hooke, J., Woodcock, R., Sixsmith, J., Wu, W., Tan, P., Li, F., Killough, B., Minchin, S., Roberts, D., Ayers, D., Bala, B., Dwyer, J., Dekker, A., Dhu, T., Hicks, A., Ip, A., Purss, M., Richards, C., Sagar, S., Trenham, C., Wang, P., and Wang, L.-W. (2017). The Australian geoscience data cube — Foundations and lessons learned. *Remote Sensing of Environment*, **202**: 276–292.

Li, J., Wu, J., and Jeon, G. (2020). GPU acceleration of clustered DPCM for lossless compression of hyperspectral images. *IEEE Transactions on Industrial Informatics*, **16**: 2906–2916.

Li, S., Dragicevic, S., Castro, F. A., Sester, M., Winter, S., Coltekin, A., Pettit, C., Jiang, B., Haworth, J., Stein, A., and Cheng, T. (2016). Geospatial big data handling theory and methods: A review and research challenges. *ISPRS Journal of Photogrammetry and Remote Sensing*, **115**: 119–133.

Liu, J., Feld, D., Xue, Y., Garcke, J., and Soddemann, T. (2015). Multicore processors and graphics processing unit accelerators for parallel retrieval of aerosol optical depth from satellite data: Implementation, performance, and energy efficiency. *IEEE Journal of Selected Topics in Applied Earth Observations and Remote Sensing*, **8**: 2306–2317.Li, W., He, C., Fu, H., and Luk, W. (2017). An FPGA-based tree crown detection approach for remote sensing images. In *2017 International Conference on Field Programmable Technology (ICFPT)*, 2017, pp. 231–234. IEEE.

Liu, D., Zhou, G., Huang, J., Zhang, R., Shu, L., Zhou, X., and Xin, C. (2019a). On-board georeferencing using FPGA-based optimized second-order polynomial equation. *Remote Sensing*, **11**(2): 124. https://doi.org/10.3390/rs11020124.

Liu, J., Xue, Y., Ren, K., Song, J., Windmill, C., and Merritt, P. (2019b). High-performance time-series quantitative retrieval from satellite images on a GPU cluster. *IEEE Journal of Selected Topics in Applied Earth Observations and Remote Sensing*, **12**: 2810–2821.

Liu, P., Di, L., Du, Q., and Wang, L. (2018). Remote sensing big data: Theory, methods and applications. *Remote Sensing*, **10**(5): 711. https://doi.org/10.3390/rs10050711.

Liu, P., Yuan, T., Ma, Y., Wang, L., Liu, D., Yue, S., and Kołodziej, J. (2014). Parallel processing of massive remote sensing images in a GPU architecture. *Computing and Informatics*, **33**: 197–217.

Liu, X. (2019). Application of cloud-based visual communication design in internet of things image. *Soft Computing*, **24**: 8041–8050.

López-Fandiño, J., B. Heras, D., Argüello, F., and Dalla Mura, M. (2017). GPU framework for change detection in multitemporal hyperspectral images. *International Journal of Parallel Programming*, **47**: 272–292.

Lukač, N., Žalik, B., Cui, S., and Datcu, M. GPU-based kernelized locality-sensitive hashing for satellite image retrieval. In *2015 IEEE International Geoscience and Remote Sensing Symposium (IGARSS)*, pp. 1468–1471. IEEE.

Luo, X., Bai, J., Chen, Y., and Tong, L. Parallel implementation of MPI-based SAR image soil moisture inversion. In *2013 IEEE International Geoscience and Remote Sensing Symposium-IGARSS*, pp. 1692–1695. IEEE.

Ma, Y., Chen, L., Liu, P., and Lu, K. (2014a). Parallel programing templates for remote sensing image processing on GPU architectures: Design and implementation. *Computing*, **98**: 7–33.

Ma, Y., Wang, L., Liu, D., Liu, P., Wang, J., and Tao, J. (2012). Generic parallel programming for massive remote sensing data processing. In *2012 IEEE International Conference on Cluster Computing*.

Ma, Y., Wang, L., Zomaya, A. Y., Chen, D., and Ranjan, R. (2014b). Task-tree based large-scale mosaicking for massive remote sensed imageries with dynamic DAG scheduling. *IEEE Transactions on Parallel and Distributed Systems*, **25**: 2126–2137.

Mullah, H.U. and Deka, B. (2018). A Fast Satellite Image Super-Resolution Technique Using Multicore Processing. In: Abraham, A., Muhuri, P., Muda, A., Gandhi, N. (Eds.), *Hybrid Intelligent Systems. HIS 2017*. Advances in Intelligent Systems and Computing, Vol. 734. Springer, Cham. https://doi.org/10.1007/978-3-319-76351-4_6.

Nico, G., Fusco, L. and Linford, J. (2003). Grid technology for the storage and processing of remote sensing data: Description of an application. *Sensors, Systems, and Next-Generation Satellites VI*, pp. 677–685. International Society for Optics and Photonics.

Ordonez, A., Arguello, F., and Heras, D. B. (2017). GPU accelerated FFT-based registration of hyperspectral scenes. *IEEE Journal of Selected Topics in Applied Earth Observations and Remote Sensing*, **10**: 4869–4878.

Osterloh, L., Pérez, C., Böhme, D., Baldasano, J. M., Böckmann, C., Schneidenbach, L., and Vicente, D. (2009). Parallel software for retrieval of aerosol distribution from LIDAR data in the framework of EARLINET-ASOS. *Computer Physics Communications*, **180**: 2095–2102.

Pakartipangi, W., Syihabuddin, B., and Darlis, D. (2015). Design of camera array interface using FPGA for nanosatellite remote sensing payload. In *2015 International Conference on Radar, Antenna, Microwave, Electronics and Telecommunications (ICRAMET)*, pp. 119–123. IEEE.

Petcu, D., Zaharie, D., Gorgan, D., Pop, F., and Tudor, D. (2007). MedioGrid: A grid-based platform for satellite image processing. In *2007 4th IEEE Workshop on Intelligent Data Acquisition and Advanced Computing Systems: Technology and Applications*, pp. 137–142. IEEE.

Peternier, A., Merryman Boncori, J. P., and Pasquali, P. (2017). Near-real-time focusing of ENVISAT ASAR Stripmap and Sentinel-1 TOPS imagery exploiting OpenCL GPGPU technology. *Remote Sensing of Environment*, **202**: 45–53.

Plaza, A., Plaza, J., and Vegas, H. (2010). Improving the performance of hyperspectral image and signal processing algorithms using parallel, distributed and specialized hardware-based systems. *Journal of Signal Processing Systems*, **61**: 293–315.

Plaza, A., Valencia, D., Plaza, J., and Chang, C. I. (2006). Parallel implementation of endmember extraction algorithms from hyperspectral data. *IEEE Geoscience and Remote Sensing Letters*, **3**: 334–338.

Pu, Y., Zhao, X., Chi, G., Zhao, S., Wang, J., Jin, Z., and Yin, J. (2019). Design and implementation of a parallel geographically weighted k-nearest neighbor classifier. *Computers & Geosciences*, **127**: 111–122.

Quesada-Barriuso, P., Blanco Heras, D., and Argüello, F. (2021). GPU accelerated waterpixel algorithm for superpixel segmentation of hyperspectral images. *The Journal of Supercomputing*, **77**: 10040–10052.

Rathore, M. M. U., Ahmad, A., Paul, A., and Wu, J. (2016). Real-time continuous feature extraction in large size satellite images. *Journal of Systems Architecture*, **64**: 122–132.

Shen, Z., Luo, J., Huang, G., Ming, D., Ma, W., and Sheng, H. (2007). Distributed computing model for processing remotely sensed images based on grid computing. *Information Sciences*, **177**: 504–518.

Sterling, T., Brodowicz, M., and Anderson, M. (2017). *High Performance Computing: Modern Systems and Practices*. Burlington: Morgan Kaufmann.

Sun, M., Liu, J., Zhang, J., Doi, K., Wang, P. S. P., Chen, Y., and Li, Q. (2007). Distributed parallel computation in DP grid system. In *MIPPR 2007: Medical Imaging, Parallel Processing of Images, and Optimization Techniques*.

Sun, Y., Liu, B., Sun, X., Wan, W., Di, K., and Liu, Z. (2014). A CPU/GPU collaborative approach to high-speed remote sensing image rectification based on RFM. In *Remote Sensing of the Environment: 18th National Symposium on Remote Sensing of China*, p. 91580F. International Society for Optics and Photonics.

Tan, K., Zhang, J., Du, Q., and Wang, X. (2015). GPU parallel implementation of support vector machines for hyperspectral image classification. *IEEE Journal of Selected Topics in Applied Earth Observations and Remote Sensing*, **8**: 4647–4656.

Tan, X., Di, L., Zhong, Y., Yao, Y., Sun, Z., and Ali, Y. (2020). Spark-based adaptive Mapreduce data processing method for remote sensing imagery. *International Journal of Remote Sensing*, **42**: 191–207.

Tie, B., Huang, F., Tao, J., Lu, J., and Qiu, D. (2018). A parallel and optimization approach for land-surface temperature retrieval on a windows-based PC cluster. *Sustainability*, **10**.

Visser, S. J., Dawood, A. S., and Williams, J. A. (2003). FPGA based satellite adaptive image compression system. *Journal of Aerospace Engineering*, **16**: 129–137.

Wang, N., Chen, F., Yu, B., and Qin, Y. (2020). Segmentation of large-scale remotely sensed images on a spark platform: A strategy for handling massive image tiles with the MapReduce model. *ISPRS Journal of Photogrammetry and Remote Sensing*, **162**: 137–147.

Wang, P., Wang, J., Chen, Y., and Ni, G. (2013a). Rapid processing of remote sensing images based on cloud computing. *Future Generation Computer Systems*, **29**: 1963–1968.

Wang, Y., Jung, Y., Supinie, T. A., and Xue, M. (2013b). A hybrid MPI–OpenMP parallel algorithm and performance analysis for an ensemble square root filter designed for multiscale observations. *Journal of Atmospheric and Oceanic Technology*, **30**: 1382–1397.

Wei, X., Liu, W., Chen, L., Ma, L., Chen, H., and Zhuang, Y. (2019). FPGA-based hybrid-type implementation of quantized neural networks for remote sensing applications. *Sensors*, **19**(4): 924. https://doi.org/10.3390/s19040924.

Weihong, W., Jiechen, C., Guomin, G., and Wei, W. (2011). Design and implementation of cluster-based platform for Remote Sensing Information Computation. In *Proceedings of 2011 International Conference on Computer Science and Network Technology*, 2011, pp. 2122–2125. IEEE.

Williams, J. A., Dawood, A. S., and Visser, S. J. (2002). FPGA-based cloud detection for real-time onboard remote sensing. In *2002 IEEE International Conference on Field-Programmable Technology, 2002. (FPT). Proceedings*, pp. 110–116. IEEE.

Wong, C. M., Bracikowski, C., Baldauf, B. K., and Havstad, S. A. (2011). Real-time 3D flash ladar imaging through GPU data processing. In *Parallel Processing for Imaging Applications*, p. 78720P. International Society for Optics and Photonics.

Wu, C., Zhang, Y. and Yang, C. (2013). Large scale satellite imagery simulations with physically based ray tracing on Tianhe-1A supercomputer. In *2013 IEEE 10th International Conference on High Performance Computing and Communications & 2013 IEEE International Conference on Embedded and Ubiquitous Computing*.

Wu, Z., Li, Y., Plaza, A., Li, J., Xiao, F., and Wei, Z. (2016). Parallel and distributed dimensionality reduction of hyperspectral data on cloud computing architectures. *IEEE Journal of Selected Topics in Applied Earth Observations and Remote Sensing*, **9**: 2270–2278.

Wu, Z., Shi, L., Li, J., Wang, Q., Sun, L., Wei, Z., Plaza, J., and Plaza, A. (2018). GPU parallel implementation of spatially adaptive hyperspectral image classification. *IEEE Journal of Selected Topics in Applied Earth Observations and Remote Sensing*, **11**: 1131–1143.

Xu, C., Du, X., Yan, Z., and Fan, X. (2020). ScienceEarth: A big data platform for remote sensing data processing. *Remote Sensing*, **12**(4): 607; https://doi.org/10.3390/rs12040607.

Xue, Y., Chen, Z., Xu, H., Ai, J., Jiang, S., Li, Y., Wang, Y., Guang, J., Mei, L., Jiao, X., He, X., and Hou, T. (2011). A high throughput geocomputing system for remote sensing quantitative retrieval and a case study. *International Journal of Applied Earth Observation and Geoinformation*, **13**: 902–911.

Xue, Y., Wan, W., and Ai, J. (2008). High performance geocomputation developments. *World SCT-Tech R&D*, **3**: 314–319.

Xue, Y., Wang, J., Wang, Y., Wu, C., and Hu, Y. (2007). Preliminary study of grid computing for remotely sensed information. *International Journal of Remote Sensing*, **26**: 3613–3630.

Yang, Y.-K., Nam, H.-O., Kim, K.-O., and Cho, S.-I. (1993). Development of supercomputer image processing software with X-Window user-interface for the processing of the remotely sensed data. *International Archives of Photogrammetry and Remote Sensing*, **29**: 235–235.

Yang, Z., Li, W., Chen, Q., Wu, S., Liu, S., and Gong, J. (2018). A scalable cyber-infrastructure and cloud computing platform for forest aboveground biomass estimation based on the Google Earth Engine. *International Journal of Digital Earth*, **12**: 995–1012.

Ye, Y., An, Y., Chen, B., Wang, J., and Zhong, Y. (2019). Land use classification from social media data and satellite imagery. *The Journal of Supercomputing*, **76**: 777–792.

Yin, Q., Wu, Y., Zhang, F., and Zhou, Y. (2020). GPU-based soil parameter parallel inversion for PolSAR data. *Remote Sensing*, **12**(3): 415. https://doi.org/10.3390/rs12030415.

Zhang, C., Di, L., Yang, Z., Lin, L., and Hao, P. (2020a). AgKit4EE: A toolkit for agricultural land use modeling of the conterminous United States based on Google Earth Engine. *Environmental Modelling & Software*, **129**: 104694. https://doi.org/10.1016/j.envsoft.2020.104694.

Zhang, N., Wei, X., Chen, H., and Liu, W. (2021). FPGA implementation for CNN-based optical remote sensing object detection. *Electronics*, **10**(3): 282. https://doi.org/10.3390/electronics10030282.

Zhang, T. and Li, J. (2015). Online task scheduling for LiDAR data preprocessing on hybrid GPU/CPU devices: A reinforcement learning approach. *IEEE Journal of Selected Topics in Applied Earth Observations and Remote Sensing*, **8**: 386–397.

Zhang, X., Wei, X., Sang, Q., Chen, H., and Xie, Y. (2020b). An efficient FPGA-based implementation for quantized remote sensing image scene classification network. *Electronics*, **9**(9). https://doi.org/10.3390/electronics9091344.

Zhen, L., Mi, W., Deren, L., and Lei, T. L. (2014). Stream model-based orthorectification in a GPU cluster environment. *IEEE Geoscience and Remote Sensing Letters*, **11**: 2115–2119.

Zheng, X., Xue, Y., Guang, J., and Liu, J. (2016). Remote sensing data processing acceleration based on multi-core processors. In *2016 IEEE International Geoscience and Remote Sensing Symposium (IGARSS)*, pp. 641–644. IEEE.

Zheng, Y., Shi, Z., He, C., and Zhang, Q. (2020). Lifting based object detection networks of remote sensing imagery for FPGA accelerator. *IEEE Access*, **8**: 200430–200439.

Zhizhin, M., Poyda, A., Velikhov, V., Novikov, A., and Polyakov, A. (2016). Data-intensive multispectral remote sensing of the nighttime Earth for environmental monitoring and emergency response. *Journal of Physics: Conference Series*, **2016**: 012029. IOP Publishing.

Zhou, H., Tang, Y., Yang, X., and Liu, H. (2007). Research on grid-enabled parallel strategies of automatic wavelet-based registration of remote-sensing images and its application in ChinaGrid. In *Fourth International Conference on Image and Graphics (ICIG 2007)*.

Zhou, H., Yang, X., Liu, H., and Tang, Y. (2006) GPGC: A grid-enabled parallel algorithm of geometric correction for remote-sensing applications. *Concurrency and Computation: Practice and Experience*, **18**: 1775–1785.

Zhong, E., Song, G., and Tang, G. (2020). *Principle, Technology and Application of Big Data Geographic Information System*. Beijing: Tsinghua University Press.

Zhu, H., Cao, Y., Zhou, Z., and Gong, M. (2012). Parallel multi-temporal remote sensing image change detection on GPU. In *2012 IEEE 26th International Parallel and Distributed Processing Symposium Workshops & PhD Forum*.

Zhu, H., Lu, L., Fan, Y., Li, P., Zhang, Q., and Jiao, L. Parallel implementation of the FLICM algorithm for SAR image change detection on intel MIC. In *2016 IEEE International Geoscience and Remote Sensing Symposium (IGARSS)*, 2016, pp. 2340–2343. IEEE.

Zou, Q., Li, G., and Yu, W. (2018). MapReduce functions to remote sensing distributed data processing-global vegetation drought monitoring as example. *Software: Practice and Experience*, **48**: 1352–1367.

Chapter 7

Data Communication

7.1 Introduction

The fundamental purpose of a communications system is the exchange of data between two parties. Figure 7.1 presents one particular example, which is communication between a workstation and a server over a public telephone network. The key elements of the model are source, transmitter, transmission system, receiver and destination (Stallings, 2000).

It is often impractical for two communicating devices to be directly, point-to-point connected. The solution to this problem is to attach each device to a communication network. Two major categories into which communications networks are traditionally classified are Wide Area Networks (WANs) and Local Area Networks (LANs). The distinction between the two, both in terms of technology and application, has become somewhat blurred in recent years, but it remains a useful way of organizing the discussion.

7.2 Wide Area Networks (WANs)

Wide area networks generally cover a large geographical area, require the crossing of public right-of-ways, and rely at least in part on circuits provided by a common carrier. Traditionally, WANs have been implemented using one of the two technologies: circuit switching or packet switching.

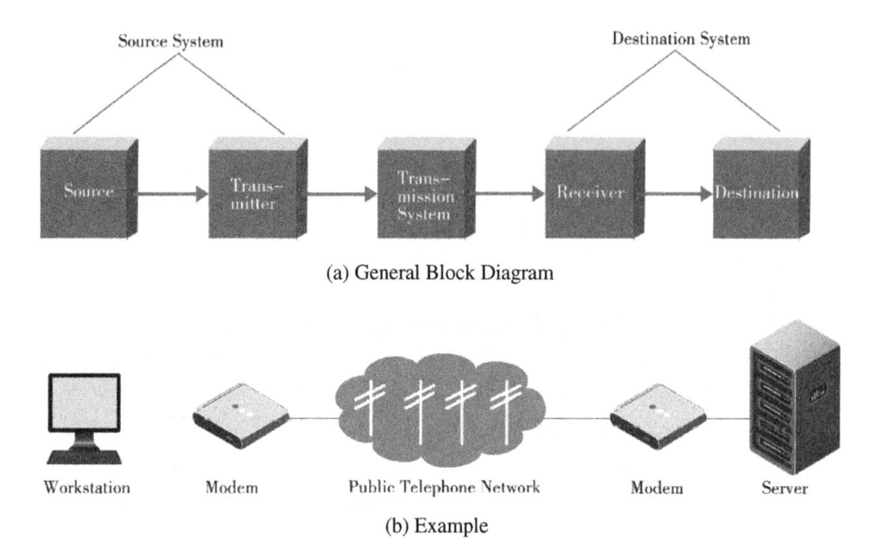

Figure 7.1. Simplified communications model.
Source: Stallings (2000).

More recently, frame relay and Asynchronous Transfer Mode (ATM) networks have assumed major roles.

It is usual for the imaging center where the remotely sensed images are generated to archive the images, and therefore long-term storage requirements are not required at the receiving sites. It may be useful to store the images online for a short time. This reduces the cost of the overall Telegeoprocessing system, as large and expensive archives are not required.

7.2.1 *Circuit switching*

In a circuit switching network, a dedicated communications path is established between two stations through the nodes of the network. That path is a connected sequence of physical links between nodes.

In a circuit-switched data network, each connection established results in a physical communication channel being set up through the network from the calling to the called subscriber equipment. This connection is then used exclusively by the two subscribers for the duration of the call. The main feature of such a connection is that it provides a fixed data rate channel, and both subscribers must operate at this rate.

7.2.2 *Packet switching*

In packet-switched data networks, all data to be transmitted are first assembled into one or more message units, called packets, by the source Data Terminal Equipment (DTE). These packets include both the source and the destination DTE network addresses. They are then passed by the source DTE to its local Packet Switching Exchange (PSE). On receipt of each packet, the PSE inspects the destination address contained in the packet. Each PSE contains a routing directory specifying the outgoing links to be used for each network address. On receipt of each packet, the PSE forwards the packet on the appropriate link at the maximum available bit rate. As each packet is received at each intermediate PSE along the route, it is forwarded on the appropriate link interspersed with other packets being forwarded on that link. At the destination PSE, determined by the destination address within the packet, the packet is finally passed to the destination DTE.

To prevent unpredictably long delays and ensure that the network has a reliably fast transit time, a maximum length is allowed for each packet. It is for this reason that a message submitted to the transport layer within the DTE may first have to be divided by the transport protocol entity into a number of smaller packet units before transmission. In turn, they will be reassembled into a single message at the destination DTE.

In packet-switched data networks, error and flow control procedures are applied on each link by the network PSEs. Consequently, the class of service provided by a packet-switched network is much higher than that provided by a circuit-switched network.

7.2.3 *Frame relay*

A high-speed packet switching protocol is used in wide area networks (WANs). Providing a granular service of up to DS3 speed (45 Mbps), it has become very popular for LAN to LAN connections across remote distances. Services are offered by all the major carriers. Frame relay is much faster than X.25 networks, the first packet-switching WAN standard, because frame relay was designed for today's reliable circuits and performs less rigorous error detection. Although X.25 was never widely used in the U.S., frame relay has become a major wide area technology. The name comes from the fact that frame relay does not do any processing of the content of the packets, rather it relays them from the input port of the switch to the output port.

Voice over frame relay enables voice to be packetized and travel over a frame relay network, often providing significant cost savings but at some sacrifice in voice quality, depending on the network configuration. In 1998, the Frame Relay Forum finalized its voice over frame relay specification. FRF.11 defines the frame formats, and FRF.12 defines the ability to divide large frames into smaller ones so that real-time voice can be interleaved with data on slower connections.

Frame relay provides permanent and switched logical connections, known as Permanent Virtual Circuits (PVCs) and Switched Virtual Circuits (SVCs). These are logical connections provisioned ahead of time (PVCs) or on demand (SVCs). The connections are identified by a Data Link Connection Identifier (DLCI) number that is significant to the local frame relay switch, which will change the number as it passes the packet on to its destination, because the receiving switch uses a different DLCI for its end of the same connection. Every DLCI requires a Committed Information Rate (CIR), which is a pledge on the part of the network to provide a certain amount of transmission capacity for the connection. CIRs are adjusted with experience.

A customer attaches to the frame relay network via a frame relay access device (FRAD) which resides on the customer's premises. The FRAD may be a separate device or software built into the router. The FRAD connects to a port on a frame relay switch on the service provider's network via an interface known as the User-to-Network Interface (UNI). This line/port is typically some multiple of 64 Kbps, and all traffic for one customer generally travels through the same port. The frame relay switches may interconnect via point-to-point lines, but they often use an ATM backbone.

The book *Frame Relay for High-Speed Networks* by Walter Goralski (1999) is a must read not only to learn about frame relay but also to learn about wide area networking in general.

7.2.4 *Asynchronous Transfer Mode (ATM)*

Asynchronous Transfer Mode (ATM) is a network technology for both local and wide area networks (LANs and WANs) that supports real-time voice and video as well as data. The topology uses switches that establish a logical circuit from end to end, which guarantees the quality of service (QoS). However, unlike telephone switches that dedicate circuits end to end, unused bandwidth in ATM's logical circuits can be appropriated when needed. For example, idle bandwidth in a videoconference circuit can be used to transfer data.

ATM is widely used as a backbone technology in carrier networks and large enterprises but never became popular as a local network (LAN) topology. ATM is highly scalable and supports transmission speeds of 1.5, 25, 100, 155, 622, 2488 and 9953 Mbps. ATM is also running as slow as 9.6 Kbps between ships at sea. An ATM switch can be added into the middle of a switch fabric to enhance total capacity, and the new switch is automatically updated using ATM's PNNI routing protocol.

7.2.4.1 *Cell switching*

ATM works by transmitting all traffic as fixed-length, 53-byte cells. This fixed unit allows very fast switches to be built because it is much faster to process a known packet size than to figure out the start and end of variable length packets. The small ATM packet also ensures that voice and video can be inserted into the stream often enough for real-time transmission.

ATM works at layer 2 of the OSI model and typically uses SONET (OC-3 and OC-12) for framing and error correction out over the wire. ATM switches convert cells to SONET frames and frames to cells at the port interface.

7.2.4.2 *Quality of Service* (*QoS*)

The ability to specify a quality of service is one of ATM's most important features, allowing voice and video to be transmitted smoothly. The following levels of service are available:

- Constant Bit Rate (CBR) guarantees bandwidth for real-time voice and video.
- Real-time variable Bit Rate (rt-VBR) supports interactive multimedia that requires minimal delays, and non-real-time variable bit rate (nrt-VBR) is used for bursty transaction traffic.
- Available Bit Rate (ABR) adjusts bandwidth according to congestion levels for LAN traffic.

- Unspecified Bit Rate (UBR) provides a best effort for non-critical data, such as file transfers.

7.2.4.3 *Integration with legacy LANs*

Network applications use protocols, such as TCP/IP, IPX, AppleTalk and DECnet, and there are tens of millions of Ethernet and Token Ring clients in existence. ATM has to coexist with these legacy protocols and networks. MPOA is an ATM standard that routes legacy protocols while preserving ATM quality of service.

LANE (LAN Emulation) interconnects legacy LANs by encapsulating Ethernet and Token Ring frames into LANE packets and then converting them into ATM cells. It supports existing protocols without changes to Ethernet and Token Ring clients but uses MPOA route servers or traditional routers for internetworking between LAN segments.

7.2.5 *Integrated Services Digital Network (ISDN) and broad ISDN*

Integrated Services Digital Network (ISDN) is an international telecommunications standard for providing a digital service from the customer's premises to the dial-up telephone network. ISDN turns one existing wire pair into two channels and four wire pairs into 23 channels for the delivery of voice, data or video. Unlike an analog modem, which converts digital signals into equivalency in audio frequencies, ISDN deals only with digital transmission. Analog telephones and fax machines are used over ISDN lines, but their signals are converted into digital by the ISDN modem.

ISDN uses 64 Kbps circuit-switched channels, called "B channels" (bearer channels), to carry voice and data. It uses a separate D channel (delta channel) for control signals. The D channel signals the carrier's voice switch to make calls, puts them on hold and activates features, such as conference calling and call forwarding. It also receives information about incoming calls, such as the identity of the caller. Since the D channel connects directly to the telephone system's SS7 signaling network, ISDN calls are dialed much faster than regular telephone calls.

ISDN's basic service is Basic Rate Interface (BRI), which is made up of two 64 Kbps B channels and one 16 Kbps D channel (2B+D). If both the channels are combined into one, called "bonding," the total data rate becomes 128 Kbps and is four and a half times the bandwidth of a V.34 modem (28.8 Kbps).

ISDN's high-speed service is Primary Rate Interface (PRI). It provides 23 B channels and one 64 Kbps D channel (23B+D), which is equivalent to the 24 channels of a T1 line. When several channels are bonded together, high data rates can be achieved. For example, it is common to bond six channels for quality videoconferencing at 384 Kbps. In Europe, PRI includes 30 B channels and one D channel, equivalent to an E1 line.

Connecting ISDN to a personal computer requires a network terminator (NT1) and an ISDN terminal adapter (TA). The NT1 plugs into the two-wire line from the telephone company with an RJ-11 connector and provides four-wire output to the TA. Within the U.S., the NT1 is typically built into the TA, but in Europe and Japan, they are separate devices.

Often called an "ISDN modem" because the device may support an analog telephone or fax machine, the TA itself is technically not a modem because it provides a digital to digital connection. External TAs plug into the serial port, while internal TAs plug into an expansion slot. Some TAs hook into the parallel port for higher speed. The TA may also include an analog modem and automatically switch between analog and digital depending on the type of call.

TAs support bonding for Internet operation, which links the channels together for higher speed, but the ISP must provide the Multilink PPP (MPPP) protocol to support this operation.

Although announced in the mid-1980s, it took more than a decade before ISDN usage became widespread. When people asked when would it arrive, the answer was ISDN: "I Still Don't Know."

7.3 Local Area Networks (LANs)

As with WANs, a LAN is a communications network that interconnects a variety of devices and provides a means for information exchange among those devices. The following are the several key distinctions between LANs and WANs (Stallings, 2000):

- The scope of the LAN is small, typically a single building or a cluster of buildings. This difference in geographic scope leads to different technical solutions, as we shall see.

- It is usually the case that the LAN is owned by the same organization that owns the attached devices. For WANs, this is less often the case or at least a significant fraction of the network assets are not owned. This has two implications. First, care must be taken in the choice of LAN because there may be a substantial capital investment (compared to dial-up or leased charges for WANs) for both purchase and maintenance. Second, the network management responsibility for a LAN falls solely on the user.
- The internal data rates of LANs are typically much greater than those of WANs.

More recently, examples of switched LANs, especially switched Ethernet LANs, have appeared. Two other prominent examples are ATM LANs, which simply use an ATM network in a local area, and Fiber Channel.

Small LANs can allow certain workstations to function as a server, allowing users access to data on another user's machine. These peer-to-peer networks are often simpler to install and manage, but dedicated servers provide better performance and can handle higher transaction volume. Multiple servers are used in large networks.

The controlling software in a LAN is the network operating system (NetWare, UNIX, or Windows NT) that resides in the server. A component part of the software resides in each client and allows the application to read and write data from the server as if it were on the local machine.

The message transfer is managed by a transport protocol, such as TCP/IP or IPX. The physical transmission of data is performed by the access method (Ethernet or Token Ring) which is implemented in the network adapters that are plugged into the machines. The actual communications path is the cable (twisted pair, coax, or optical fiber) that interconnects each network adapter.

Wireless LAN is a local area network that transmits over the air typically in an unlicensed frequency, such as the 2.4 GHz band. A wireless LAN does not require lining up devices for the line of sight transmission like IrDA. Wireless access points (base stations) are connected to an Ethernet hub or server and transmit a radio frequency over an area of several hundred to a thousand feet which can penetrate walls and other non-metal barriers. Roaming users can be handed off from one access point to another like a cellular phone system. Laptops use wireless modems that plug into an existing Ethernet port or that are self-contained on PC cards, while stand-alone desktops and servers use plug-in cards (ISA or PCI).

There have been numerous proprietary products on the market for home and office, but Proxim's OpenAir and the IEEE 802.11 are two major standards for which numerous products are available. Bluetooth and HomeRF are home and small office technologies that are expected to proliferate in the 2000–2002 time frame. Such systems have a more limited range and do not support roaming. Small wireless LANs are sometimes called "personal area networks" (PANs) since one of their primary uses is to serve an individual connecting a laptop or PDA to a desktop machine.

Wireless LANs function like cellphone systems. Each access point is a base station that transmits over a radius of several hundred feet. In systems designed for office use, users can seamlessly roam between access points without dropping the connection. Figure 7.2 shows a wireless LAN organization.

Typical GIS operations are grouped by class and load on the communication network (Table 7.1). Table 7.2 shows the comparison of speed by service type (Kirton, 1995).

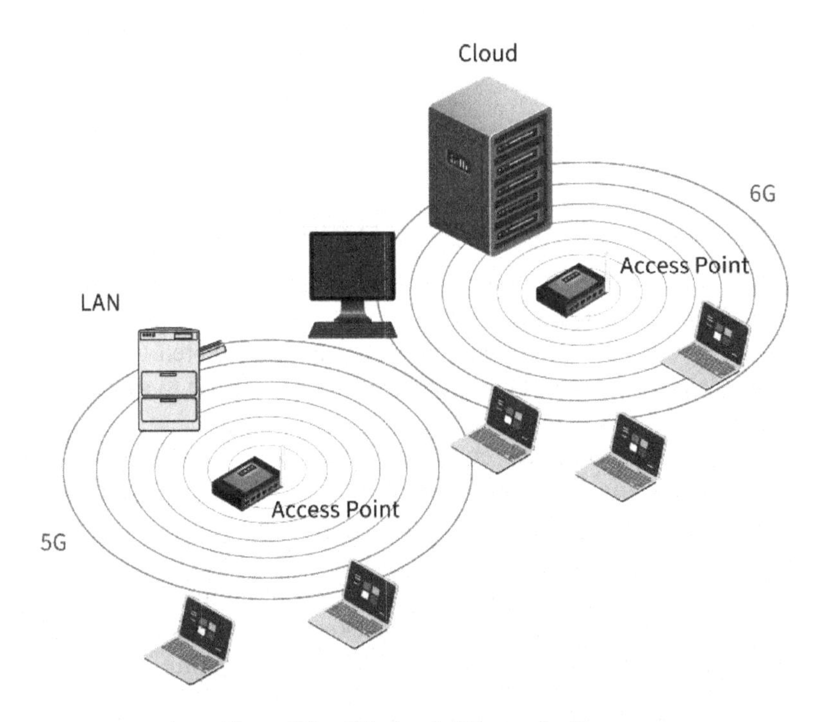

Figure 7.2. Wireless LAN organization.

Table 7.1. Potential GIS-related usage of broadband communications under different information management approaches.

	Degree of centralization in...		Possible GIS-related usage of high-speed communications network
Approach	Processing	Storage	
1	Decentralized	Decentralized	'Browsing' of remote files prior to retrieval; high-speed file transfer between sites
2	Centralized	Centralized (all data are stored on a central file server)	Any activity is possible through network file sharing, including copying of selected files to local workspace and processing, editing, analyzing and displaying of centrally stored files
3	Centralized	Decentralized (Local storage possible; only corporate data stored on central server)	Query and display of centrally stored 'corporate data'; transfer of selected files to local workspace for processing
4	Centralized	Centralized	Display and edit (i.e., re-drawing) processes would comprise the major load on the communication network

Source: Zwart and Coleman (1995).

7.4 Protocols and Protocol Architecture

The exchange of information between computers for the purpose of cooperative action is generally referred to as computer communications. Similarly, when two or more computers are interconnected via a communication network, the set of computer stations is referred to as a computer network.

In discussing computer communications and computer networks, two concepts are paramount: Protocols and Computer-communications architecture, or protocol architecture. A protocol is used for communication between entities in different systems. The terms entity and system are used in a very general sense. Examples of entities are user application programs, file transfer packages, database management systems, electronic mail facilities, and terminals. Examples of systems

Table 7.2. Comparison of speed by service type.

	Speed		Transmission time (Times are for full screens without compression)				
Service type	Speed range	Speed of example	Text screen 10 kbits	G3 fax A4 page 240 kbits	Color image PC screen 2.5 Mbits	Color image high resolution 24 Mbits	Remotely sensed image 48 Mbits
PSTN modem	300–9600 bps	2400 bps	4.2 s	1.7 min	17.3 min	2.8 hr	5.6 hr
PSPDN	2.4–56 Kbps	9.6 bps	1.0 s	25 s	4.3 min	42.7 min	1.4 hr
ISDN	64 Kbps	64 Kbps	0.16 s	3.8 s	39 s	6.3 min	12.5 min
ISDN PR or Frame Relay	64 Kbps–2 Mbps	1 Mbps	10 ms	0.24 s	2.5 s	24 s	48 s
Man	0–45 Mbps	10 Mbps	1 ms	24 ms	0.25 s	2.4 s	4.8 s
B-ISDN	0–120 Mbps payload	50 Mbps	0.2 ms	4.8 ms	50 ms	0.48 s	0.96 s
155 Mbps (ATM)	0–155 Mbps	100 Mbps	0.1 ms	2.4 ms	25 ms	0.24 s	0.48 s

Notes: bps is bits per second; Kbps is Kilobits per second; Mbps is Million bits per second; kbits is Kilobits per second

Source: Kirton (1995).

are computers, terminals, and remote sensors. The key elements of a protocol are Syntax (includes such things as data format and signal levels), Semantics (includes control information for coordination and error handling), and Timing (includes speed matching and sequencing).

7.4.1 *The TCP/IP architecture*

Transmission Control Protocol/Internet Protocol (TCP/IP) is a communications protocol developed under a contract from the U.S. Department of Defense to internetwork dissimilar systems. Invented by Vinton Cerf and Bob Kahn, this *de facto* UNIX standard is the protocol of the Internet and has become the global standard for communications.

TCP provides transport functions, which ensures that the total amount of bytes sent is received correctly at the other end. UDP, which is part of the TCP/IP suite, is an alternate transport that does not guarantee delivery. It is widely used for real-time voice and video transmissions where erroneous packets are not retransmitted.

TCP/IP is a routable protocol, and the IP part of TCP/IP provides this capability. In a routable protocol, all messages contain not only the address of the destination station but also the address of a destination network. This allows TCP/IP messages to be sent to multiple networks (subnets) within an organization or around the world, hence its use on the worldwide Internet.

IP accepts packets from TCP or UDP, adds its own header and delivers a "datagram" to the data link layer protocol. It may also break the packet into fragments to support the maximum transmission unit (MTU) of the network.

Every client and server in a TCP/IP network requires an IP address, which is either permanently assigned or dynamically assigned at startup.

TCP/IP is composed of two parts: Transmission Control Protocol (TCP) and Internet Protocol (IP). TCP is a connection-oriented protocol that passes its data to IP, which is a connectionless one. TCP sets up a connection at both ends and guarantees reliable delivery of the full message sent. TCP tests for errors and requests retransmission if necessary because IP does not.

An alternative protocol to TCP within the TCP/IP suite is User Datagram Protocol (UDP), which does not guarantee delivery. Like IP, it is also connectionless but very useful for real-time voice and video, where it doesn't matter if a few packets get lost because it's too late to worry about it if they do.

Although TCP/IP has five layers, it is contrasted to the Open Systems Interconnection (OSI) seven-layer model (see Figure 6.3) because OSI serves as a universal reference. Once thought to become the worldwide communications standard, OSI gave way to TCP/IP. However, it has become the teaching model for all communications networks, so all efforts of the OSI committee were not in vain.

Perhaps the simplest reference ever written on the subject is *An Introduction to TCP/IP* by John Davidson (1988), published by Springer-Verlag. Although written in 1988 and only 100 pages, it is the easiest read on the subject you will ever find.

The Bibles for TCP/IP have been *Internetworking with TCP/IP*, Volumes I, II and III, by Douglas E. Comer. Written in 1991, Volume I covers the essentials from soups to nuts, published by Prentice Hall.

7.4.2 *The Open Systems Interconnection (OSI) model*

Open Systems Interconnection is an ISO standard for worldwide communications that defines a framework for implementing protocols in seven layers (Figure 7.3). Control is passed from one layer to the next, starting at the application layer in one station, proceeding to the bottom layer, over the channel to the next station and back up the hierarchy.

At one time, most vendors agreed to support OSI in one form or another, but OSI was too loosely defined and proprietary standards were too entrenched. Except for the OSI-compliant X.400 and X.500 e-mail and directory standards, which are widely used, what was once thought to become the universal communications standard now serves as the teaching model for all other protocols.

Most of the functionality in the OSI model exists in all communications systems although two or three OSI layers may be incorporated into one.

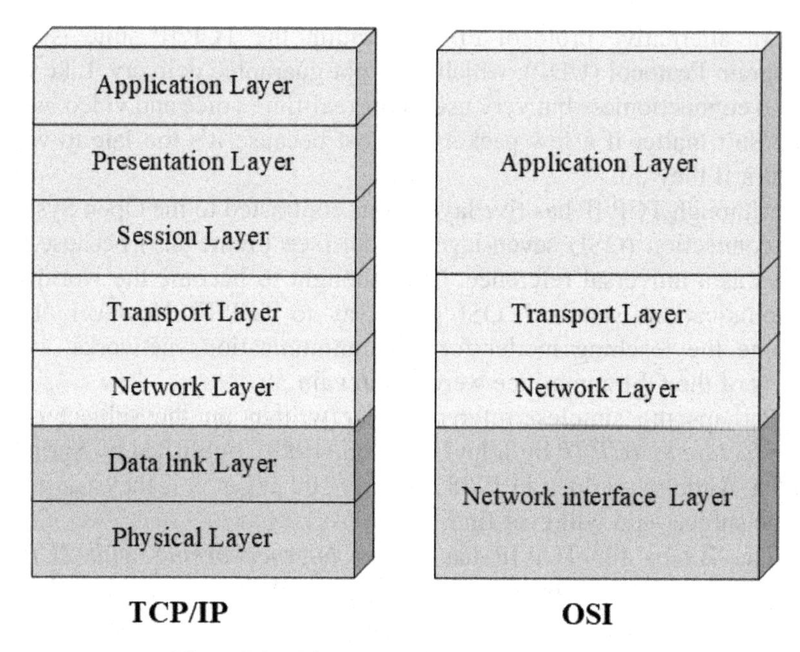

Figure 7.3. The OSI model and TCP/IP model.

7.5 Transmission Media

In a data transmission system, the transmission medium is the physical path between the transmitter and receiver. Transmission media can be classified as guided or unguided. In considering the design of data transmission systems, key concerns are data rate and distance. A number of design factors relating to the transmission medium and the signal determine the data rate and distance, which are as follows:

- Bandwidth;
- transmission impairments;
- interference;
- number of receivers.

Figure 7.4 depicts the electromagnetic spectrum and indicates the frequencies at which various guided media and unguided transmission techniques operate. Each frequency range has a band designator and each range of frequencies behaves differently and performs different functions.

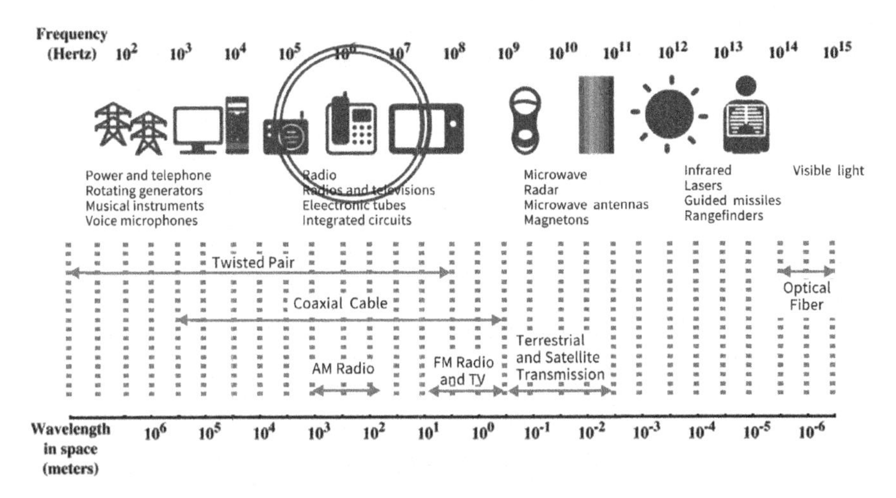

Figure 7.4. Electromagnetic spectrum for telecommunications.

Source: Stallings (2000).

Table 7.3. International band designators.

Designation	Frequency range
ELF: extremely low frequency	3–30 Hz
SLF: super low frequency	30–300 Hz
ULF: ultra low frequency	300–3,000 Hz
VLF: very low frequency	3–30 kHz
LF: low frequency	30–300 kHz
MF: medium frequency	300–3,000 kHz
HF: high frequency	3–30 MHz
VHF: very high frequency	30–300 MHz
UHF: ultra high frequency	300–3,000 MHz
SHF: super high frequency	3–30 GHz
EHF: extremely high frequency	30–300 GHz

The descriptive designation of international designators are listed in Table 7.3.

For communications purposes, the usable frequency spectrum now extends from about 3 Hz, through 300 GHz, and up to about 100 THz, where research on laser communications is taking place. This frequency spectrum is shared by civil, government, and military users of all nations

Table 7.4. Point-to-point transmission characteristics of guided media.

	Frequency range	Typical attenuation	Typical delay (μs/km)	Repeater spacing (km)
Twisted pair (with loading)	0–3.5 kHz	0.2 dB/km @ 1 kHz	50	2
Twisted pair (multipair cables)	0–1 MHz	3 dB/km @ 1 kHz	5	2
Coaxial cable	0–500 MHz	7 dB/km @ 10 MHz	4	1–9
Optical fiber	180–370 THz	0.2–0.5 dB/km	5	40

Source: Glover and Grant (1998).

according to International Telecommunications Union (ITU) radio regulations.

7.5.1 *Guided transmission media*

The three guided media commonly used for data transmission are twisted pair, coaxial cable, and optical fiber. Table 7.4 indicates the characteristics typical for the common guided media for long-distance point-to-point applications.

7.5.2 *Wireless transmission*

Furthermore, wireless Internet permits scientists and referring decision makers to view images on their desktop personal computer (PC), laptop computer, and WAP phone anywhere and anytime, without having to go to a particular department where fixed Internet connections are available. This additional networking tool is currently in use at a small scale and is in testing status.

7.5.2.1 *Broadcast radio*

In addition to its uses for the public broadcast of radio and television programs and for private communication with devices like portable phones, electromagnetic radiation can be used to transmit computer data.

Informally, a network that uses electromagnetic radio waves is said to operate at radio frequency, and the transmissions are referred to as RF transmission. Radio waves are easy to generate, can travel long distances, and can penetrate buildings easily, so they are widely used for communication, both indoors and outdoors. Radio waves also are omnidirectional, meaning that they travel in all directions from the source, so the transmitter and receiver do not have to be carefully aligned physically.

Unlike networks that use wires or optical fibers, networks using RF transmissions do not require a direct physical connection between computers. Instead, each participating computer attaches to an antenna, which can both transmit and receive RF. Physically, the antennas used with RF networks may be large or small, depending on the range desired. For example, an antenna designed to propagate signals several miles across a town may consist of a metal pole approximately two meters long that is mounted vertically on top of a building. An antenna designed to permit communication within a building may be small enough to fit inside a portable computer (e.g., less than 20 centimeters).

The RF range extends from about 10,000 Hz to over 300,000,000,000 (300 GHz). For convenience, the Federal Communications Commission (FCC) has divided the RF spectrum into different bands. These frequency bands, their uses, their characteristics, and their advantages and disadvantages are addressed in detail later in this chapter. As weather personnel, we're mainly concerned with radio operations in the HF, VHF, UHF, and sometimes SHF frequency bands for our communications. However, with the added emphasis on satellite communications in our career field, expect more involvement with communications equipment operating through the EHF frequency band.

7.5.2.2 *Terrestrial and satellite microwave*

The most common type of microwave antenna is the parabolic "dish". A typical size is about 3 meters in diameter. Microwave antennas are usually located at substantial heights above ground level to extend the range between antennas and to be able to transmit over intervening obstacles. To achieve long-distance transmissions, a series of microwave relay towers are used, and point-to-point microwave links are strung together over the desired distance.

An increasingly common use of microwaves is for short point-to-point links between buildings. This can be used for closed-circuit TV or as a data link between local area networks. Short-haul microwaves can also be used for the so-called bypass application. A business can establish a microwave link to a long-distance telecommunications facility in the same city, bypassing the local telephone company. The most common bands for long-haul telecommunications are the 4 GHz to 6 GHz bands. Higher-frequency microwave is being used for short point-to-point links between buildings; typically, the 22 GHz band is used.

Although radio transmissions do not bend around the surface of the earth, RF technology can be combined with satellites to provide communication across longer distances. The satellite contains a transponder that consists of a radio receiver and transmitter. The transponder accepts an incoming radio transmission, amplifies it, and transmits the amplified signal back toward the ground at a slightly different angle than it arrived. Since placing a communication satellite in orbit is expensive, a single satellite usually contains multiple transponders that operate independently (typically 6 to 12). Each transponder uses a different radio frequency (i.e., channel), making it possible for multiple communications to proceed simultaneously. Furthermore, because a single satellite channel can be shared, it can serve many customers. There are two common configurations for satellite communication. In the first, the satellite is being used to provide a point-to-point link between two distant ground-based antennas. In the second, the satellite provides communications between one ground-based transmitter and a number of ground-based receivers.

The optimum frequency range for satellite transmission is in the range of 1–10 GHz. Most satellites providing point-to-point service today use a frequency bandwidth in the range of 5.925–6.425 GHz for transmission from Earth to a satellite (uplink) and a bandwidth in the range of 4.7–4.2 GHz for transmission from a satellite to Earth (downlink). This combination is referred to as the 4/6 GHz band. The other two bands are 12/14 GHz band (uplink: 14–14.5 GHz; downlink: 11.7–14.2 GHz) and 19/29 GHz band (uplink: 27.5–31.0 GHz; downlink: 17.7–21.2 GHz).

Due to the long distance involved, there is a propagation delay of about a quarter second from transmission from one Earth station to reception by another Earth station.

7.5.2.3 *Infrared*

Infrared Data Association (www.irda.org) was founded in 1993 and is dedicated to developing standards for wireless, infrared transmission systems between computers. With IrDA ports, a laptop or PDA can exchange data with a desktop computer or use a printer without a cable connection. IrDA requires line-of-sight transmission like a TV remote control. IrDA products began to appear in 1995. The LaserJet 5P was one of the first printers with a built-in IrDA port.

IrDA is comprised of the IrDA Serial IR (IrDA-SIR) physical layer, which provides a half-duplex connection of up to 115.2 Kbps. This speed allows the use of a low-cost UART chip; however, higher non-UART, high-speed extensions up to 4 Mbps for Fast Infrared (FIR) have also been defined. IrDA uses the Infrared Link Access Protocol (IrLAP), an adaptation of HDLC, as its data link protocol. The Infrared Link Management Protocol (IrLMP) is also used to provide a mechanism for handshaking and multiplexing two or more different data streams simultaneously.

7.5.2.4 *Bluetooth*

A wireless personal area network (PAN) technology from the Bluetooth Special Interest Group (www.bluetooth.com) was founded in 1998 by Ericsson, IBM, Intel, Nokia and Toshiba. Bluetooth is an open standard for short-range transmission of digital voice and data between mobile devices (laptops, PDAs, and phones) and desktop devices. It supports point-to-point and multipoint applications.

Bluetooth provides up to 720 Kbps data transfer within a range of 10 meters and up to 100 meters with a power boost. Unlike IrDA, which requires that devices be aimed at each other (line of sight), Bluetooth uses omnidirectional radio waves that can transmit through walls and other non-metal barriers. Bluetooth transmits in the unlicensed 2.4 GHz band and uses a frequency hopping spread spectrum technique that changes its signal 1600 times per second. If there is interference from other devices, the transmission does not stop, but its speed is downgraded.

The name Bluetooth comes from King Harald Blatan (Bluetooth) of Denmark. In the 10th century, he began to Christianize the country. Ericsson (Scandinavian company) was the first to develop this specification.

7.6 Data Communications Interfacing

A variety of standards for interfacing exists. The most widely used interfacing is one that is specified in the ITU-T standard: V.24. In fact, this standard specifies only the functional and procedural aspects of the interface; V.24 references other standards for the electrical and mechanical aspects. It does not describe the connectors or pin assignments; those are defined in ISO 2110. In the U.S., EIA-232 incorporates the control signal definition of V.24, the electrical characteristics of V.28 and the connector and pin assignments defined in ISO 2110.

V.24 defines the functions of all circuits for the Recommended Standard-232 (RS-232) interface. RS-232 is a TIA/EIA standard for serial transmission between computers and peripheral devices (modem or mouse). Using a 25-pin DB-25 or 9-pin DB-9 connector, its normal cable limitation of 50 feet can be extended to several hundred feet with high-quality cable.

RS-232 defines the purpose and signal timing for each of the 25 lines; however, many applications use less than a dozen. RS-232 transmits positive voltage for a 0 bit, negative voltage for a 1. In 1984, this interface was officially renamed TIA/EIA-232-E standard (E is the current revision, 1991) although most people still call it RS-232.

Pin Settings for Plug
(Reverse order for socket.)

1 2 3 4 5 6 7 8 9 10 11 12 13
14 15 16 17 18 19 20 21 22 23 24 25

1 2 3 4 5
6 7 8 9

The wide variety of functions available with V.24/EIA-232 are provided by the use of a large number of interchange circuits. This is a rather expensive way to achieve results. Another most important data communication interfacing standard is the ISDN physical interface.

In ISDN terminology, a physical connection is made between the terminal equipment (TE) and network-terminating equipment (NT). The physical connection, defined in ISO 8877, specifies that the NT and TE cables shall terminate in matching connectors that provide eight contacts.

7.7 Communications Security

Network security problems can be divided roughly into four closely intertwined areas: secrecy, authentication, non-repudiation, and integrity control (Tanenbaum, 2003). Secretly, also called confidentiality, has to do with keeping information out of the hands of unauthorized users. Authentication deals with determining whom you are talking to before revealing sensitive information or entering into a business deal. Non-repudiation deals with signatures. Finally, how can you be sure that a message you received was really the one sent and not something that a malicious adversary modified in transit or concocted?

Network security threats fall into two categories. Passive threats, sometimes referred to as eavesdropping, involve attempts by an attacker to obtain information relating to a communication. Active threats involve some modification of the transmitted data or the creation of false transmissions.

7.7.1 *Encryption*

By far the most important automated tool for network and communications security is encryption. The messages to be encrypted, known as the plaintext, are transformed by a function that is parameterized by a key. The output of the encryption process, known as the ciphertext, is then transmitted, often by messenger or radio.

Conventional encryption is also referred to as symmetric encryption or single-key encryption. A conventional encryption scheme has five ingredients: plaintext, encryption algorithm, secret key, ciphertext and decryption algorithm. With conventional encryption, two parties share a single encryption/decryption key. The principal challenge with conventional encryption is the distribution and protection of the keys. Since the real secrecy is in the key, its length is a major design issue. The longer the key, the higher the work factor the cryptanalyst has to deal with. The work factor for breaking the system by exhaustive search of the key space is exponential in the key length. For routine commercial use, at least 128 bits should be used. To keep major governments at bay, keys of at least 256 bits, preferably more, are needed.

The most commonly used conventional encryption algorithms are block ciphers. A block cipher processes the plaintext input in fixed-size blocks and produces a block of ciphertext of equal size for each plaintext block. The two most important conventional algorithms, both of which are block ciphers, are the Data Encryption Standard (DES) and Triple DEA (TDEA). DES was developed by IBM as its official standard for unclassified information. U.S. government, National Bureau of Standards, now the National Institute of Standards and Technology (NIST), adopted it as Federal Information Processing Standard 46 (FIPS PUB46). It is no longer secure in its original form, but in a modified form it is still useful. The plaintext must be 64 bits in length and the key is 56 bits in length; longer plaintext amounts are processed in 64-bit blocks.

TDEA was first proposed by Tuchman (1979) and first standardized for use in financial applications in ANSI standard X9.17 in 1985. TDEA was incorporated as part of the DES in 1999, with the publication of FIPS PUB 46-3. TDEA uses three keys and three executions of the DES algorithm (Stallings, 2003). With three distinct keys, TDEA has an effective key length of 168 bits. FIPS 46-3 also allows for the use of two keys (Tanenbaum, 2003).

Public-key encryption was first publicly proposed by Diffie and Hellman in 1976 (Diffie and Hellman, 1976). For one thing, public-key algorithms are based on mathematical functions rather than on simple operations on bit patterns. A public-key encryption scheme involves two keys: one for encryption and a paired key for decryption. One of the keys is kept private by the party that generated the key pair, and the other is made public. The encryption and decryption keys are different, and the decryption key could not feasibly be derived from the encryption key.

The RSA public-key encryption algorithm is one of the first public-key schemes, which was developed by a group at MIT (Rivest *et al.*, 1978). It has survived all attempts to break it for more than a quarter of a century and is considered very strong. Much practical security is based on it. Its major disadvantage is that it requires a key of at least 1024 bits for good security (versus 128 bits for conventional encryption algorithms), which makes it quite slow. RSA is a block cipher in which the plaintext and ciphertext are integers between 0 and n-1 for some n.

Conventional encryption and public-key encryption are often combined in secure networking applications. Conventional encryption is used to encrypt transmitted data, using a one-time or short-term session key. The session key can be distributed by a trusted key distribution center or transmitted in encrypted form using public-key encryption. Public-key encryption is also used to create digital signatures, which can authenticate the source of transmitted messages.

A security enhancement used with both Ipv4 and Ipv6, called IPSec, provides both confidentiality and authentication mechanisms. The IPSec specification is quite complex and covers numerous documents. The most important of these, issued in November of 1998, are Request for Comments (RFCs) 2401, 2402, 2406, and 2408 (www.ietf.org).

7.7.2 *Firewalls*

In addition to the danger of information leaking out, there is also a danger of information leaking in. In particular, viruses, worms, and other digital pests can breach security, destroy valuable data, and waste large amounts of administrators' time trying to clean up the mess they leave. Consequently, mechanisms are needed to keep "good" bits in and "bad" bits out. One method is to use IPSec. This approach protects data in transit between secure sites. However, IPSec does nothing to keep digital pests and intruders from getting onto LANs.

Firewall is a method for implementing security policies designed to keep a network secure from intruders. It can be a single router that filters out unwanted packets or may comprise a combination of routers and servers each performing some type of firewall processing. Firewalls are widely used to give users secure access to the Internet as well as to separate a company's public Web server from its internal network. Firewalls are also used to keep internal network segments secure; for example, the accounting network might be vulnerable to snooping from within the enterprise. In practice, many firewalls have default settings that provide little or no security unless specific policies are implemented by trained personnel.

Firewalls installed to protect entire networks are typically implemented in hardware; however, software firewalls are also available to protect individual workstations from attack. Following are the techniques used in combination to provide firewall protection:

- Packet Filter: Blocks traffic based on a specific Web address (IP address) or type of application (e-mail, ftp, or Web), which is specified by port number. Also known as a "screening router."
- Proxy Server: Serves as a relay between two networks, breaking the connection between the two. Also typically caches Web pages.

- Network Address Translation (NAT): Allows one IP address, which is shown to the outside world, to refer to many IP addresses internally; one on each client station. Performs the translation back and forth.
- Stateful Inspection: Tracks the transaction to ensure that inbound packets were requested by the user. Generally can examine multiple layers of the protocol stack, including the data, if required, so blocking can be made at any layer or depth.

7.7.3 *Wireless security*

Wireless data are coming and in a big way. Wireless is a snooper's dream come true: free data without having to do any work. While protecting voice communications from interception is a mission of newer wireless networks from a physical transmission perspective, these developments do not preclude the necessity of session-based encryption with Wireless Transport Layer Security (WTLS) specification. Standards like global system for mobile (GSM) and Cellular Digital Packet Data (CDPD) provide encryption that has proven fairly robust, but past history does not guarantee future success.

A large risk with a wireless network is that of eavesdropping. Most of the security problems can be traced to the manufacturers of wireless base stations (access points) trying to make their products user-friendly. If the user takes the device out of the box and plugs it into the electrical power socket, it begins operating immediately — nearly always with no security at all, blurting secrets to everyone within radio range. Wireless terminals use radio communication, which passes right over the firewall in both directions. The range of 802.11 networks is often a few hundred meters, so anyone who wants to spy on a LAN can simply drive into the range area and leave an 802.11-enabled notebook computer to record everything it hears. It therefore goes without saying that security is even more important for wireless systems than for wired ones. In this section, we will look at some ways wireless networks handle security. Some additional information can be found in Nichols and Lekkas's book (Nichols and Lekkas, 2002).

7.7.3.1 *802.11 security*

802.11 is a family of IEEE standards for wireless LANs which was first introduced in 1997. The first standard was 802.11b, which specifies from

1 to 11 Mbps in the unlicensed 2.4 GHz band using direct sequence spread spectrum (DSSS) technology.

Using the orthogonal FDM (OFDM) transmission method, there are two subsequent standards that provide from 6 to 54 Mbps: 802.11a transmits in the higher 5 GHz frequency range and is not backward compatible with the slower 802.11b; 802.11g works in the same range and is compatible. The 11a and 11b standards are endorsed and branded as "Wi-Fi" by the Wi-Fi Alliance.

An 802.11 system works in two modes. In "infrastructure mode," wireless devices communicate to a wired LAN via access points. Each access point and its wireless devices are known as a Basic Service Set (BSS). An Extended Service Set (ESS) is two or more BSSs in the same subnet. In "*ad hoc* mode," also known as "peer-to-peer mode," wireless devices can communicate with each other directly and do not use an access point. This is an Independent BSS (IBSS).

The speed of 802.11 systems is distance-dependent. The farther the remote device is from the base station, the lower the speed.

The 802.11 standard prescribes a data link-level security protocol called Wired Equivalent Privacy (WEP), which is designed to make the security of a wireless LAN as good as that of a wired LAN. When 802.11 security is enabled, each station has a secret key shared with the base station. WEP encryption uses a stream cipher based on the RC4 algorithm. While this approach looks good at the first glance, a method for breaking it has already been published (Borisov *et al.*, 2001; Fluhrer *et al.*, 2001).

7.7.3.2 *Bluetooth security*

Bluetooth has a considerably shorter range than 802.11, so it cannot be attacked from the radio range, but security is still an issue here. Bluetooth has three security modes, ranging from nothing at all to full data encryption and integrity control. Bluetooth provides security in multilayers (Tanenbaum, 2003). Another security issue is that Bluetooth authenticates only devices, not users. However, Bluetooth also implements security in the upper layers, so even in the event of a breach of link-level security, some security may remain, especially for applications that require a PIN code to be entered manually from some kind of keyboard to complete the transaction.

7.7.3.3 *WAP 2.0 security*

WAP is partially a rebirth of an earlier standards effort, the handheld device markup language. WAP aims to have its transport protocol closely parallel TCP/IP, without carrying forward that protocol's overhead, which makes it ill-suited for wireless. WAP is intended to operate over any of the different wireless transmission technologies, such as Cellular Digital Packet Data (CDPD), Code Division Multiple Access (CDMA) and global system for mobile (GSM). The WAP standard related to security is called the Wireless Transport Layer Security (WTLS) specification. WTLS specification is based upon its TCP/IP counterpart: Secure Sockets Layer.

WAP2.0 largely uses standard protocols in all layers. Security is no exception. Since it is IP-based, it supports full use of IPSec in the network layer. Since WAP 2.0 is based on well-known standards, there is a chance that its security services, in particular, privacy, authentication, integrity control, and non-repudiation, may fare better than 802.11 and Bluetooth security.

7.7.3.4 *Satellite spread spectrum communication security*

The disadvantages of the openness of satellite communication channels, such as poor concealment and weak anti-interference ability, can be overcome by using spread spectrum technology. Therefore, spread spectrum communication is mainly used for covert communication and anti-interference military communication. Spread spectrum mainly includes direct sequence spectrum spreading (DSSS), frequency hopping (FH), hopping time and linear frequency modulation.

In a DSSS system, each symbol is represented by a pseudo-random sequence of length N, which can extend the frequency band of its signal by *n*-fold, and the receiver adopts the same sequence to carry out correlation reception and de-expansion. Therefore, the signal-to-noise ratio after de-expansion can be increased to n-fold before de-expansion, and *n*-fold de-expansion processing gain can be obtained. Therefore, it can communicate under the condition of a low signal-to-noise ratio, which can make the communication signal have strong concealment and make the system have high interference tolerance. Based on DSSS, the transparent transponder of a GEO satellite can be used to form an overlapping communication system with strong concealment. DSSS signals with extremely low power spectral density are overlapped on other strong signals that are being communicated and then carry out communication with a low bit rate to be high concealment.

In FH communication, the sender will modulate the signal in a wide frequency range according to a certain secret frequency hopping pattern, and the receiver will use the local oscillation of the same jump to carry out orthogonal down conversion and then return to the zero intermediate frequency signal for baseband demodulation, symbol decision and decoding. Therefore, FH is easier to extend the signal spectrum to a wider frequency range than DSSS and can obtain higher processing gain. As long as the frequency hopping range is wide enough and the hopping speed is fast enough, and the satellite multibeam antenna technology is combined to avoid possible interference from space, the security of communication will be fully guaranteed.

7.7.3.5 *Fiber optic communication security*

Optical fiber communication is one of the most promising communication technologies in the world. Optical fiber communication systems are composed of optical transceivers, repeaters and optical cables, which have the advantages of wide frequency band, large communication capacity, small attenuation, anti-interference ability, strong security, low cost and so on. The principle of optical fiber communication is as follows: the sending end transmitted information (such as voice) into electrical signals and then modulated to the laser beam emitted by the laser so that the intensity of light changes with the amplitude (frequency) of electrical signals and sent out through the optical fiber. At the receiving end, the detector converts the optical signal into an electrical signal and restores the original information after demodulation.

In the aspect of communication security and anti-interference, the measures commonly adopted include the following: spread spectrum communication, frequency hopping communication, digital communication, scattering communication, laser communication, burst communication, and radio silence and management. Although these measures can play a certain role in confidentiality and anti-interference, they cannot be compared with optical fiber communication. Therefore, in the future commercial, military and other confidential communications, optical fiber communication is undoubtedly one of the most secure and reliable confidential anti-interference communication methods.

7.8 4G and 5G

7.8.1 *4G communications*

4G communication technology is the fourth generation of mobile information systems, which is a better improvement than 3G technology. Compared with 3G communication technology, a greater advantage of 4G is to combine the WLAN technology and 3G communication technology so that the image transmission speed is faster and the quality of the transmitted image becomes better. The application of 4G communication technology in intelligent communication equipment allows users to access the Internet faster, and the speed can be up to 100 M, which is much faster than 3G communication technology.

4G communication technology has the following characteristics:

(1) Significantly improve the communication speed: Compared with the previous 3G communication technology, the biggest advantage of 4G communication technology is to significantly improve the communication speed so that users have a better user experience. At the same time, it also promotes the development of communication technology. The development of communication technology is a long process, the first generation of mobile information systems only has the voice system, and the development of the second generation of mobile information system communication speed is only 10 Kb/s. In the development of 3G communication technology, the speed is not a qualitative leap, which only reached 2 Mb/s. All these have become an obstacle to the development of the communication industry, but the emergence of 4G communication technology is obviously a qualitative leap in communication speed.
(2) More intelligent communication technology: 4G communication technology has achieved intelligent operation to a large extent compared to the previous mobile information system. According to the different commands, the intelligent 4G communication technology can make a more accurate response to search out the data analysis, processing and finishing to transfer to the user's mobile phone. As a communication tool that people increasingly cannot do without, a 4G mobile phone greatly facilitates people's life.
(3) Improvements in compatibility: The degree of cooperation between software and hardware is usually referred to as compatibility. If there

are fewer conflicts between software and hardware, compatibility will be improved; if there are more conflicts, compatibility will be reduced. The emergence of 4G communication technology greatly improves the performance of compatibility and reduces the conflicts between software and hardware in the working process, which largely avoids the occurrence of program errors. One performance of 4G communication technology that greatly improves compatibility is that we will rarely encounter various errors, such as delaying and exiting abnormally, which often occur before, making people feel more smooth and fluent in the process of using communication equipment.

In a word, 4G technology not only makes our life easier to a greater extent but also promotes the development of the communication industry. With its higher performance, faster running speed and more intelligent operation, 4G technology has been very popular in many fields.

7.8.2 *5G communications*

5G is a new generation of mobile communication systems designed to meet the needs of mobile communication after 2020. According to the laws of the development of mobile communication, 5G will have ultra high spectrum efficiency and energy efficiency. Compared with 4G mobile communications, the transmission rate and resource utilization of 5G will increase by an order of magnitude or more, and it will also be significantly improved in the wireless coverage performance, transmission delay, system security and user experience. 5G mobile communications will be closely combined with other wireless mobile communication technology and generate a new ubiquitous mobile information network to meet the development of mobile Internet traffic in the next 10 years. The application field of 5G mobile communication systems will be further extended, and the ability to support the huge amounts of sensing equipment and machine to machine (M2M) communications will become one of the important indicators of the system design. The future 5G system should also have sufficient flexibility and intelligent capabilities in the network, such as self-perception and self-adjustment so that it can cope with the unpredictable and rapid changes in the future mobile information society.

5G has become a research hotspot in the field of mobile communication at home and abroad. In early 2013, the EU launched the Mobile and Wireless Communications Capabilities for the 2020 Information Society (METIS) project for 5G research in the 7th Framework Plan. At present, countries around the world are conducting extensive discussions on the development vision, application requirements, candidate frequency bands, key technical indicators and enabling technologies of 5G, striving to reach a consensus before and after the 2015 World Radio Conference and start the relevant standardization process after 2016.

The vigorous development of mobile Internet is the main driving force of 5G mobile communication. Mobile Internet will be the basic service platform for various emerging services in the future. Various services of the existing fixed Internet will be provided to users more and more wirelessly, and the wide application of cloud computing and background services will put forward higher transmission quality and system capacity requirements for 5G mobile communication systems. The main development goal of 5G mobile communication systems will be to closely connect with other wireless mobile communication technologies and provide all-around basic business capabilities for the rapid development of mobile Internet. According to the current preliminary estimates of the industry, the future improvement of wireless mobile network service capability, including 5G, will be carried out simultaneously in the following three dimensions:

(1) By introducing new wireless transmission technology, the resource utilization will be increased by more than 10 times on the basis of 4G.
(2) By introducing new architecture (such as ultra dense cell structure) and deeper intelligent capability, the throughput of the whole system is increased by about 25 times.
(3) Further tap new frequency resources (such as high-frequency band, millimeter wave and visible light) to expand the frequency resources of wireless mobile communication by about four times in the future.

The iconic key technologies of 5G mobile communication are mainly reflected in ultra efficient wireless transmission technology and high-density wireless network technology. Among them, the wireless transmission technology based on large-scale MIMO may improve the spectrum efficiency and power efficiency by another order of magnitude on the basis of 4G. The main bottleneck of this technology towards practicality is high-dimensional channel modeling with estimation and complexity control. Full duplex technology may open up a new pattern of spectrum utilization in a new generation of mobile communication. Network coordination and interference management will be the core and key issues to improve the capacity of high-density wireless networks.

References

Borisov, N., Goldberg, I., and Wagner, D. (2001). Intercepting mobile communications: The insecurity of 802.11. In *Proceeding 7th International Conference on Mobile Computing and Networking*, ACM, pp. 18–188.

Diffie, W. and Hellman, M. (1976). New directions in cryptography. *IEEE Transactions on Information Theory*, November 1976.

Fluhrer, S., Mantin, I., and Shamir, A. (2001). Weakness in the key scheduling algorithm of RC4. In *Proceeding 8th Annual Workshop on Selected Areas in Cryptography*.

Glover, I. and Grant, P. (1998). *Digital Communications*. Upper Saddle River, NJ: Prentice Hall.

Goralski, W. J. (1999). *Frame Relay for High-Speed Networks*. 1st edn. John Wiley & Sons, Inc.605 Third Ave. New York, NY, United States.

Nichols, R. K. and Lekkas, P. C. (2002). *Wireless Security*. New York: McGraw-Hill.

Stallings, W. (2000). *Data and Computer Communications*. New Jersey: Prentice Hall International, Inc.

Tuchman, W. (1979). Hellman presents no shortcut solutions to DES. *IEEE Spectrum*.

Chapter 8

A Prototype Telegeoprocessing System

A real-time system is defined as a system that is required to satisfy a set of timing constraints as well as ensure the semantic correctness of the system. For example, in image processing that involves screen updating for viewing continuous motion, deadlines are of the order of 30 ms. In this case, response time is defined as the time between input and the corresponding display. Although often it is desirable that the system be fast, real-time systems differ from non-real-time systems in requiring timely behavior and predictability (Halang and Stoyenko, 1991). In practice, there are many imaging applications that are time-critical and computationally intensive (e.g., satellite image-data processing and target tracking). Also, many image processing applications lend themselves to parallelization and hence parallel architectures. The following requirements for a telegeoprocessing system have been compiled based on the experiences described above. Depending on the specific application, multimedia systems for telegeoprocessing will require different combinations of the hardware and/or software components described under each of the following categories.

8.1 Media Acquisition

Images may be obtained from a number of sources. Unfortunately, many digitized images cannot be directly transmitted and may require a redundant step of further digitization, as the specific equipment vendors may not generate images in an accepted electronic format for transmission.

The ideal solution would be for all the imaging data to be acquired digitally and for it to be generated in the OpenGIS standard format. This avoids the unnecessary steps of producing hard-copy film and subsequent re-digitization.

Digital video may be either available directly (e.g., digital bitstreams from CD-ROM and digital video disks) or acquired through the combination of an analog video source, such as a video camera and a video digitizer. Audio may be obtained using an audio digitizer.

8.2 Media Storage

Video and audio clips and images require temporary or permanent local storage. This storage can be provided through either magnetic and/or magneto-optical (MO) drives. If the telegeoprocessing system is incorporated within a GIS, new incoming and old comparison images can be stored permanently in the GIS archive database and accessed by the telegeoprocessing system as needed.

It is usual for the imaging center where the remotely sensed images are generated to archive the images, and therefore long-term storage requirements are not required at the receiving sites. It may be useful to store the images online for a short time. This reduces the cost of the overall telegeoprocessing system, as large and expensive archives are not required.

8.3 Compression/Decompression

The necessity for image compression is readily appreciated from Table 8.1. Potentially large file sizes are reduced to manageable levels for image transmission and storage. Instead of images potentially taking hours to be transmitted, images can arrive in a matter of seconds or minutes. This has two significant advantages. The obvious advantage is a major cost saving on telecommunications charges, particularly for international telegeoprocessing. However, telegeoprocessing performed domestically can also benefit as images could be transmitted via lower cost conventional telephone lines (28 kilobits per second — kbps) rather than investing in lines with higher bandwidth which are more expensive. Perhaps a less obvious advantage is that any telegeoprocessing system is likely to be far more effective with a rapid turnaround from image

Table 8.1. Effect on transmission times (min) with compression for images (25 MB).

	Bandwidth				
Compression ratio	14.4 Kbps**	20.8 Kbps	128 Kbps	1.544 Mbps (T1)	155 Mbps (ATM)
1:1	231.5*	115.7	25.9	2.2	0.02
2:1	115.7	3	12.9	1.1	0.01
10:1	23.1	11.5	2.6	0.22	0.002
20:1	11.5	5.7	1.3	0.11	0.001
30:1	7.7	3.8	0.88	0.07	0.0006
60:1	3.8	1.9	0.44	0.04	0.0003

Notes: *All transmission times in minutes.
**Kilobits per second.

acquisition to map products. Without compression, a series of images may take many hours to be transmitted, severely limiting the ability of the decision makers to return a decision within a reasonable time frame. Obviously, emergency studies would necessitate fast turnaround times, which without image compression would require relatively expensive bandwidth.

Image compression is rapidly becoming standard in any GIS or remote sensing system due to the potentially enormous image file sizes required for transmission and for storage/archive requirements (Table 8.1). Digital compression techniques are used widely in the telecommunications field but have come somewhat late to the geoprocessing field. Compression techniques are either "lossless", where there is no loss of image data, or "lossy", where redundant information in an image file can be discarded. Compression of remotely sensed images has been historically reversible or "lossless", limiting compression ratios to between 2:1 and 4:1. Unfortunately, "lossless" techniques are not very useful for telegeoprocessing. "Lossy" techniques are now gaining widespread acceptance as an effective means of reducing file sizes while maintaining image quality. The technical explanation of how lossy compression functions is beyond the scope of this paper, but simply stated, redundant information which is not crucial to the interpretative image can be discarded, thereby potentially significantly reducing image file sizes. Lossy compression is based either on the JPEG standard or wavelet compression techniques. Although the JPEG standard is supported by the OpenGIS standard, it permits compression only up to 20:1 compression, which may not be high enough for large volume use. At higher compression ratios using JPEG algorithms, block artifacts tend to appear, causing appreciable image degradation (Erickson *et al.*, 1998). Wavelet compression algorithms are much more robust at higher compression levels, and several studies have demonstrated effective image "virtually lossless" quality above 20:1 compression ratios (Xue and Rees, 1999), which produce statistically identical interpretative results compared with using the original images without any lossy compression. Although wavelet compression techniques appear to perform better than JPEG methods, wavelet compression is not currently supported by OpenGIS standards. However, it is expected that the next version of OpenGIS standards will address wavelet compression issues. If this kind of compression is properly used and does not require much additional time for compression and decompression, it can significantly reduce the communications bandwidth, storage requirements, and overall delay in telegeoprocessing systems.

Some compression can be accomplished in software, depending on the main CPU of the telegeoprocessing system. To achieve the necessary image and video compression for telegeoprocessing at real-time rates, a dedicated compression/decompression board based on programmable DSPs such as special compression chipsets from a company — Analog Analysis — is normally required. Chau *et al.* (1995) discuss the development of a real-time image encoder and apply a co-design methodology that considers both hardware and software options. After investigation of possible hardware/software implementations of a real-time image encoder, the authors then select a good design which meets the system specification.

The accepted international standard for videoconferencing is H.320 (ITU-T Rec., 1994) which includes support for video (H.261) and audio (G.722 and G.728) compression/decompression sharing (T.120). H.320 is designed to work over the range of ISDN connections (from 64 Kbps to 1.92 Mbps). There exist other compression standards which support higher-quality video and have correspondingly higher bandwidth requirements. Motion JPEG is a "symmetric" coder/decoder (codec) (it requires roughly the same amount of computation to encode and decode frames), which eliminates intraframe redundancy. Better compression ratios could be obtained by utilizing both interframe and intraframe redundancies. These algorithms require "asymmetric" codecs (which take significantly more computation to encode than to decode the frames due to motion estimation between successive video frames) such as MPEG-1 and MPEG-2 (MPEG-1 CD, 1991, MPEG-2 CD, 1994). MPEG-1 is mainly designed to compress video at higher bit rates. MPEG-2 is more flexible and supports various combinations of levels and profiles from VHS-quality up to high-definition television (HDTV)-quality video.

Guo and Forstert (1993) describe a real-time lossless data compression algorithm for both individual band and multispectral remote sensing images. The proposed technique uses prediction to reduce redundancy, and the classical Huffman coding scheme to reduce the average code length. Benz *et al.* (1995) present novel algorithms for synthetic aperture radar (SAR) raw data compression. The algorithms lead to high resolution and good SNR.

8.4 Image Transmission

Although in an ideal world, images could be transferred using the highest bandwidth available (i.e., ATM at 155 Mbps), this is either not possible due to telecommunication infrastructure restrictions or feasible due to prohibitively high costs. Fortunately, due to image compression, extremely high bandwidth is not necessary. Relatively large files can be reduced to manageable levels. Table 8.2 shows the comparison of speed by service type (Kirton, 1995).

For low volume telegeoprocessing, simple telephone lines (28.8 Kbps) may suffice, particularly if compression is used. However, should more than a few studies require transmission, then an integrated services digital network (ISDN) is recommended (128 Kbps). Ultimately, the bandwidth required should be tailored to the particular needs of the telegeoprocessing service. High volume telegeoprocessing will require higher bandwidths (and higher compression ratios). The bandwidth required may also depend on the types of studies to be transmitted and the expected peak activity.

The advantages of ATM are higher bandwidths and statistical multiplexing of small packets (cells) with guaranteed bandwidth and minimal latency and jitter (Handel *et al.*, 1993). According to ACTS (1994), the major benefit of ATM is the unification of traffic types (voice, data, and video). The ATM technology acts as a multiservice integrator allowing combination of isochronous traffic (with real-time constraints) and connection-oriented traffic, based on variable data rate and connectionless service, over a single set of physical links. These characteristics are well suited for supporting multimedia services, which are already attracting a wide range of users and service providers. ATM is also distance, protocol, and speed independent. It can be implemented through the entire network, offering LAN-WAN integration and simplifying network management and operation.

Unlike ISDN, the range of bandwidths supported by ATM is sufficient for the entire range of telegeoprocessing applications, including MPEG-2 video streams. A large image transfer of 250 Mb would require 1.6 s at 155 Mbps without compression, ignoring network overhead. With 20:1 compression and ignoring the time necessary to compress and decompress the images, this transfer would require only 0.08 s at 155 Mbps. In addition, ATM also offers “bandwidth on demand”, which allows a connection to deliver a higher bandwidth only when it is needed. The disadvantages of using ATM for telegeoprocessing are the high costs and scarcity of ATM equipment and telecommunications lines, especially to rural areas. However, ATM equipment and line availability have been increasing steadily and are expected to improve considerably in the future, and costs are expected to decrease as the size of the ATM market and user acceptance increase. It should be noted that ATM is one of the technologies being evaluated by telecommunications and interactive television companies for providing video dial tone services to the home.

If the intended images for transmission are in a standard format, then image transmission is relatively straightforward. OpenGIS compatible images offer flexibility and are readily distributed by most new vendor GISs. However non-standard images will require proprietary software to permit the digital files to be recognized by the telecommunication system that is being used. Either way, it is possible for the full 12-bit dataset to be transmitted, allowing window level and width manipulation at the receiving end. Analog film that requires digitization will only contain 8-bit information and window level and width control cannot be performed. The use of proprietary digital files (non-standard) involves significant disadvantages in that they cannot generally be received into other telegeoprocessing systems. This important fact should be borne in mind when either a GIS or telegeoprocessing system is being purchased.

As expected, the Internet is offering a convenient method of transmitting images at low cost, high bandwidth, and wide accessibility. The Internet is already being used to transmit images between the different sites. The cost of transmission is significantly reduced as only a “local call” is required to connect to the local Internet service provider (ISP) which transmits the images anywhere in the world. However, this "local call" could be the "weak link in the chain" unless relatively high bandwidth (i.e., ISDN) to the ISP provider is used. Issues of data confidentiality are of course paramount to any image transmission, particularly for the Internet. For telegeoprocessing, the image compression process can encrypt the data so that the study is unrecognizable unless it is decompressed with proprietary software. Data confidentiality issues will inevitably require stricter standards to prevent casual theft of data. It is likely, however, that the Internet will become the standard telecommunication system for telegeoprocessing due to cost and flexibility.

Furthermore, wireless Internet permits scientists and referring decision makers to view images on their desktop personal computer (PC), laptop computer, WAP phone, etc. anywhere and anytime, without having to go to a particular department where fixed Internet connections are available. This additional networking tool is currently in use at small scale and testing status.

8.5 Image and Video Processing and Interpretation

Image processing requirements for telegeoprocessing applications can be derived from duplicating the functionality available to corresponding tasks performed in the scientific environment without telegeoprocessing. Basic image manipulation functions such as 90-degree rotations and horizontal and vertical flips are essential to correct the errors in image acquisition and ensure that images can be presented to the scientists and decision makers in the way in which they are accustomed to viewing them. This is particularly important in telegeoprocessing. Zooming and panning are necessitated by the limited spatial resolution of cathode ray tubes (CRTs). Real-time window/level (brightness and contrast adjustment) is required by the need to examine images interactively with more than 8b/pixel by adjusting the range (window) and the center position (level) in the wide input dynamic range. In the case of video, manipulation functions such as play, record, pause, and rewind are important for simulating the VCR environment often used.

Often image segmentation is performed to better describe the dynamics of the Earth's ecosystems. Le Moigne and Tilton (1995) present region-based and edge-based segmentation approaches for this type of analysis. Haverkamp *et al.* (1995) describe a classification scheme for sea ice thickness based on dynamic local thresholding from SAR data. Research work has been carried out in optimizing classification algorithms. For example, Ochieng-Oglla *et al.* (1995) present an efficient iterative restoration algorithm and its application to the problem of real-time processing of broadband synthetic aperture sonar (SAS) data.

Appell *et al.* (1994) describe a real-time port information system developed for Tampa Bay, Florida. The system measures various parameters (e.g., water level, current, and weather) from around the bay and provides the data in real-time via a packet-modem-controlled telemetry system. Preston *et al.* (1994) describe a configurable multispectral imaging and classification system that support automated real-time target/background discrimination. The proposed system consists of a passive, multispectral imaging electro-optical sensor and real-time digital data collection with a data fusion image processor.

Region of Interest (ROI) live video is mainly for users interested in or hot areas, such as large open-air plazas, tourist attractions, and important transportation hubs for live video in order to meet the needs of users to view the real-time state of the region. The processing type of ROI video live broadcast is video location perception. Satellite on-orbit processing process mainly includes video sequence image positioning, video image stabilization, and video compression. The expected processing effect is to live video through domestic ground stations or relay satellite ground stations, overlooking the real-time dynamics of the selected hot spots.

5G, 6G, and artificial intelligence technology provide important support for reliable low latency transmission and intelligent research of intelligent remote sensing satellites. In the new era, the real-time service technology of intelligent remote sensing satellites is studied. The application mode of remote sensing image data is changed from "post-processing and distribution" to "on-orbit processing and real-time transmission".

For example, change detection mainly detects single frame image data, including two satellite-to-ground collaborative processing modes: one is the upshot mode. Target detection is a real-time intelligent detection of typical static/dynamic targets such as aircraft and ships under the condition of limited on-board resources. Before the task instructions are upshot, historical reference images and other data are upshot through the high-speed uplink channel. Second, the storage mode, for a specific area twice or more imaging, the previous shot image processing and storage on the satellite, with the latter shot image change detection. For the two application modes of compression and non-compression of on-board raw data, there are generally two target detection techniques. Under the mode of raw data compression, the compressed domain target information extraction technology is used to mine the image target feature information contained in the compressed domain by using the sparse dictionary generated in the image compression process. Since the target information extraction and image compression in compressed domain adopt the same set of data sparse and feature modeling methods, the target feature information can be directly extracted from compressed stream, so as to effectively save on-board computing and storage resources. The target ROI fast filtering technology is used in the original data without compression mode, which mainly filters the massive original data obtained by the sensor on orbit quickly, and obtains a small amount of data ROI collection that only contains the target and the surrounding neighborhood, so as to save on-board storage resources and reduce the amount of data transmitted by satellite and earth.

The satellite on-orbit processing flow of change detection task mainly includes image positioning, image registration, change detection, and image compression. The expected processing effect is that the change information in the target area can be extracted through the domestic ground station or relay satellite ground station, and the topological semantic information such as change range and type can be obtained. Moving target extraction and tracking is mainly carried out for video data, mainly for but not limited to the extraction and tracking of moving aircraft, ships, and large vehicles. The satellite on-orbit processing flow of moving target extraction and tracking mainly includes video sequence image positioning, video stabilization, moving target detection, moving target tracking, and video compression. The expected processing effect is to extract the trajectory and velocity information of a moving target in the observation area through the domestic ground station or the relay satellite ground station and track its motion state. The leap from GB-level massive remote sensing data to KB-level effective remote sensing information is realized, and the breakthrough of remote sensing satellite from professional user-oriented service to mass mobile terminal service is completed.

8.6 User Interface

The user interface should be graphical due to the fact that much of the information being shared in a telegeoprocessing system is inherently graphical. One or more high-resolution displays, a keyboard, a pointing device such as a mouse, and a window manager such as Microsoft Windows or UNIX X-window make up a basic telegeoprocessing user interface. These are all in addition to the multimedia devices (camera, microphones, and speakers) required by the application. In the future, a telegeoprocessing interface needs to become "simpler". A single display, a single primary input device such as a reliable voice recognition unit for commands and report writing or a virtual reality glove and automation of the most common tasks such as communications connection and disconnection and bandwidth allocation could significantly reduce the "information overload" of telegeoprocessing interfaces and allow the user to focus on the tasks at hand.

8.7 Network Interfaces

Required network interfaces for telegeoprocessing can range from low to high bandwidth, depending on the application. Low bandwidth interfaces should support multiple links because low bandwidth connections are often combined together to provide the bandwidth necessary for telegeoprocessing. These interfaces include V.34 plain old telephone service (POTS) connections, 56 Kbps dedicated or frame reply connections, ISDN connections from 64 Kbps to 1.92 Mbps, and fractional and full T1 interfaces which provide up to 1.54 Mbps per connection. Higher bandwidth interfaces include TAXI (100 Mbps) and SONET (155 Mbps and higher). Furthermore, interfaces between the wide area network and local area networks (e.g., Ethernet and fiber distributed data interface (FDDI)) will be required to allow the scientists and decision makers to access imaging devices and other geographic information systems regardless of where they are located.

In a telegeoprocessing system using special hardware for compression, it is highly desirable for the network interface to be tightly integrated with the compression hardware. This decreases network latency and reduces the processing load on the host. Bus mastering support by both the network interface and the compression hardware in a single-bus configuration could provide good performance, assuming other bus masters (including the host CPU) release control of the bus quickly. A better alternative would be to use a connection separate from the host bus, such as a second bus, or a high-speed serial connection, such as IEEE P1394 (Firewire) (IEEE P1394/D7, 1994).

8.8 Network Protocols

Support for standard networking protocols is critical to telegeoprocessing systems in order to meet the performance and interoperability requirements. ATM is the preferred internetworking protocol between telegeoprocessing systems based on the bandwidth and quality of service requirements of geoscience video and imaging applications (Kawarasaki, 1991). Wide area networking connections at rates lower than T1 require ISDN or similar protocols. In locations where ISDN is not available, POTS could be used to support H.324 videoconferencing; this is a derivative of H.320 which includes support for communications at up to 64 Kbps. In addition, support for Transmission Control Protocol/Internet Protocol (TCP/IP) is most likely a requirement on both the LAN and WAN interfaces for access to geoscience data and other resources on remote local area networks.

Standards for supporting real-time audio and video services over most networking protocols, including ATM, are still being defined. Unlike H.320 (which was designed specifically for ISDN), most video compressing methods have not been designed for a specific networking model. For instance, MPEG-2 depends on a constant channel delay to receive timing information properly (MPEG-2 CD, 1994). ATM guarantees a maximum variance in delay but not a constant delay. High variation implies larger buffering for delay-sensitive traffic such as voice or video. There is also a significant amount of redundancy in timing, multiplexing, and error detection and recovery information between the MPEG-2 transport stream and the ATM adaptation layer. The MPEG-2 quantization scale, which determines the compression ratio, does not automatically adjust to changes in the available network bandwidth. Therefore, although it is possible to carry MPEG-2 over ATM, the two standards are currently not well integrated. The ATM Forum is in the process of establishing a standard for better support of MPEG-2 over ATM (ATM Forum, 1995). Ways to guarantee a certain video or audio quality in the presence of congestion and other properties of real-world networks need to be investigated further.

8.9 Communications Requirements

Typical GIS operations are grouped by class and load on the communication network (Table 8.2). The cumulative communications requirements of a telegeoprocessing system vary from one application to another.

Table 8.2. Comparison of speed by service type.

Service type	Speed: Speed range	Speed: Speed of example	Transmission time (Times are for full screens without compression): Text screen 10 kbits	G3 fax A4 page 240 kbits	Color image PC screen 2.5 Mbits	Color image high resolution 24 Mbits	Remote sensed image 48 Mbits
PSTN modem	300–9600 bps	2400 bps	4.2 s	1.7 min	17.3 min	2.8 hr	5.6 hr
PSPDN	2.4–56 Kbps	9.6 bps	1.0 s	25 s	4.3 min	42.7 min	1.4 hr
ISDN	64 Kbps	64 Kbps	0.16 s	3.8 s	39 s	6.3 min	12.5 min
ISDN PR or Frame Relay	64 Kbps–2 Mbps	1 Mbps	10 ms	0.24 s	2.5 s	24 s	48 s
Man	0–45 Mbps	10 Mbps	1 ms	24 ms	0.25 s	2.4 s	4.8 s
B-ISDN	0–120 Mbps payload	50 Mbps	0.2 ms	4.8 ms	50 ms	0.48 s	0.96 s
155 Mbps (ATM)	0–155 Mbps	100 Mbps	0.1 ms	2.4 ms	25 ms	0.24 s	0.48 s

Approach	Degree of centralization: Processing	Degree of centralization: Storage	Possible GIS-related usage of high-speed communications network
1	Decentralized	Decentralized	• "Browsing" of remote files prior to retrieval; • High-speed file transfer between sites
2	Centralized	Centralized (All data stored on central file server)	Any activities possible through network file sharing, including • Copying of selected files to local workspace; • Processing, edit, analysis, and display of centrally stored files.

Table 8.2. (*Continued*)

Approach	Degree of centralization: Processing	Degree of centralization: Storage	Possible GIS-related usage of high-speed communications network
3	Centralized	Decentralized (Local storage possible; only corporate data stored on central server)	• Query and display of centrally stored "corporate data"; • Transfer of selected files to local workspace for processing.
4	Centralized	Centralized	• Display and edit (i.e., re-drawing) processes would comprise the major load on the communication network

Notes: bps is bits per second; Kbps is Kilobits per second; Mbps is Million bits per second; kbits is Kilobits per second
Source: Kirton (1995).

The following discussion describes the general communication requirements of each bitstream common to telegeoprocessing systems.

8.9.1 *Video*

Telegeoprocessing systems could require up to two or three simultaneous video bitstreams: two low-rate bitstreams for teleconferencing with an optional high-rate bitstream for analysis video. H.261 (64 Kbps to 1.92 Mbps), H263 (15–34 Kbps), or MPEG-1 (1.2–2 Mbps) is suitable for each teleconferencing video stream, while better quality compression algorithms such as MPEG-1 or MPEG-2 at 3–15 Mbps might be needed for the analysis video stream. In all cases, latency (a measure of the time taken to transmit data) and jitter (a measure of variance in the network arrival rate) of the real-time video streams are critical and should be minimized.

8.9.2 *Audio*

In addition to video, telegeoprocessing systems require up to two or three simultaneous audio bitstreams: two low-rate bitstreams for teleconferencing, with an optional high-rate bitstream for analysis audio. For example, basic G.711 at 56 Kbps is noisy and considered unacceptable. However, G.722 (48–64 Kbps), G.723 (5–6 Kbps), G.726 (32 Kbps), and G.728 (16 Kbps) are more acceptable for teleconferencing. Diagnostic audio streams could require CD-quality audio such as MPEG-1 layer 2 audio (32–256 Kbps) or Dolby AC-3 (96–768 Kbps). Teleconferencing audio needs to be synchronized with the video streams, and, as with video, latency and jitter of the real-time audio bitstreams should be minimized.

8.9.3 *Images*

Images are generally transmitted unidirectionally, so a single image transfer stream is sufficient. Image transfers are high-volume (e.g., 10–250 Mb/transfer) and bursty (low mean transfer rate with high maximal transfer rates). Therefore, the image transfer stream could be normally disabled and enabled dynamically as needed. The bandwidth required is dependent on the types of images supported and the acceptable image delay. For example, a typical remote sensing image such as AVHRR data is at a resolution of 2048 pixels by 2048 pixels and 10 b/pixel with each pixel stored in 2 bytes. If the maximum acceptable delay is 10 s/image, a bandwidth of at least 7 Mbps would be required to transfer the image within this limit, ignoring network overhead and traffic. Unlike video and audio streams, latency is less critical to image transfers and jitter is irrelevant.

8.9.4 *Records*

Electronic record transfers are unidirectional transfers principally text information. Text is compact and requires little bandwidth. Even an abnormally large record of 100 typewritten pages would represent only 3 MB of uncompressed information. However, frequent additions of non-text information such as scanned handwritten records and other graphics records to the electronic record could significantly increase the size of the record and the required bandwidth. Records are also fairly static. Most of the records could be transferred prior to the telegeoprocessing session. As with image transfers, bandwidth should be allocated to record transfer during a telegeoprocessing session only as needed, latency is not critical, and jitter is irrelevant.

8.10 On-orbit Intelligent Remote Sensing Data Processing

Intelligent remote sensing satellite is originated from on-orbit processing technology. Intelligent remote sensing satellite is a new type of remote sensing satellite that is reconfigurable and extensible. It has the ability for autonomous task planning, remote sensing data processing, information extraction, and inter-satellite and inter-satellite data information transmission. It can adapt to the needs of rapid, accurate, and flexible remote sensing data acquisition and information product production. It has the characteristics of design which gradually shifts from hardware definition to software definition, the function gradually changed from single to integrated, task planning gradually shifts from ground planning to on-board intelligent planning, the mission planning of intelligent remote sensing satellite adopts the form of on-board intelligent planning, the application mode gradually shifts from product-driven to task-driven and event-aware, the management mode has gradually shifted from the separation of remote control to the integration of remote control services, and intelligence level gradually shifts from single-star intelligence to multistar collaborative group intelligence.

Since the 1990s, researchers have carried out relevant research. On-board mission planning of intelligent remote sensing satellites first requires users to submit observation mission requirements, including the available time window information of ground mobile stations or relay satellites, which are fed to intelligent remote sensing satellites through inter-satellite communication links. On-board mission planning system receives these mission requirements information, according to the transit information and cloud information of the observation target, under the constraints of the observation target, the satellite resource constraints, and the relationship between the target and the satellite; the intelligent solution algorithm is used to solve the task planning scheme with the largest comprehensive benefit value, automatically generate task instructions, and obtain data independently. On-board intelligent mission planning can also combine the results of on-board intelligent image processing to determine whether the required observation target information data are obtained according to user requirements and decide whether to continue data acquisition in the next working cycle. Intelligent mission planning greatly reduces the dependence on ground control, reduces the amount of data injected into the mission instructions of complex functional satellites, and truly makes the results of intelligent image processing respond in time.

In the field of remote sensing image on-orbit real-time processing, research organizations and commercial companies around the world have carried out a large number of on-board real-time processing algorithms, which promote the development of the whole process of remote sensing data real-time acquisition, intelligent processing, sparse compression, and real-time distribution. Table 8.3 shows the real-time on-orbit processing applications of foreign remote sensing satellites.

Table 8.3. Applications of remote sensing satellites in orbit real-time processing.

Satellite name	Country	Launch year	Satellite processing	Technical approach
EO-1	America	2000	Emergency detection, feature detection, and anomaly detection	MongooseV processor
BIRD	Germany	2001	Multitype remote sensing image preprocessing, real-time multispectral classification on the satellite	TMS320C40 floating-point dsp, FPGA, and NI1000 network protocol processor
NEMO	America	2003	Hyperspectral adaptive compression	SHARC DSP
X-SAT	Singapore	2006	Automatic elimination of invalid data	Virtex FPGA, StrongARM
Pleiades-1/2	France	2011/2012	Radiation correction, geometric correction, and image compression	MVP Modular Processor Based on FPGA

Source: Wang and Yang (2019).

In order to transmit real-time data collected by satellite video to the ground and meet users' more intuitive observation needs of moving targets in remote sensing video, it is necessary to realize on-orbit real-time coding and transmission of key information under the condition of limited on-board computing resources and transmission bandwidth. The intelligent compression of satellite video is to compress the original data to the range that the satellite-ground bandwidth can withstand through an efficient intelligent compression algorithm under the condition of limited computing resources on the satellite. It meets the requirements of satellite-ground transmission in three aspects: compression efficiency, coding frame rate, and compression quality. The moving target detection of satellite video needs to detect the moving target in the compression domain and mark the detection results in the decoder with a wireframe, so as to meet the needs of real-time processing of satellite video on the satellite and make full use of the intermediate results in the compression process to meet the observation needs of users. In terms of video data compression, Skybox of the United States has applied the coding technology of general video such as H. 264 to Skysat satellite video compression.

In addition, intelligent remote sensing satellite on-orbit processing algorithm also includes change detection, video stabilization, semantic segmentation, scene classification, morphological housing index calculation, control point matching, sequence image space–time fusion, and three-dimensional reconstruction. These algorithms are combined and applied together to provide basic guarantee for real-time on-orbit processing and intelligent service of remote sensing images.

The on-board high-precision geometric positioning technology of an optical image is based on the rigorous imaging geometric model. On the basis of geometric calibration, according to the storage capacity and processing capacity of the on-board environment, the optical satellite rigorous imaging collinear equation model suitable for the on-board real-time processing unit is constructed. Combined with the ground internal and external calibration system, the parameters of the on-board imaging model are corrected and optimized to realize the on-board high-precision real-time positioning. According to the coverage range of the target area, based on the global digital elevation model (DEM) data, the image of the target area is corrected by block indirect method, and the level-2 latitude and longitude projection products with geographical coordinates are produced.

For the relative radiation correction of optical images, first, it is necessary to analyze the various factors affecting the photoelectric response of the probe element. Based on the maximum *a posteriori* probability theory, the on-orbit adaptive system radiation correction model is constructed by introducing constraints and weighting ideas. Then, the intelligent screening model of samples is constructed by using the calibration data or real-time remote sensing data, and the calculation accuracy of radiometric calibration parameters is optimized by using the strategy of incremental statistics. Finally, based on the ground radiometric calibration results, the calibration lookup table is updated to the star, and the relative radiometric correction is carried out on the star according to the lookup table to obtain the corrected image. The key to optical image relative radiometric correction is to determine the calibration model and obtain the calibration parameters. For the limited storage space on the satellite, it is necessary to determine the relative radiometric correction algorithm suitable for the satellite based on the principle of minimizing the calibration lookup table and ensuring the correction quality.

Cloud detection is to achieve accurate, fast, and efficient detection and extraction of cloud areas in images. The focus is to distinguish cloud from high-reflective targets such as ice and snow and cloud-like targets in some flat areas in waters, rivers, and urban scenes. By obtaining accurate cloud region extraction results, accurate cloud cover information is provided for subsequent ROI region extraction and multitemporal image change detection. There are many types of cloud detection algorithms. Due to the different principles of each type of algorithm, the computational resources, memory resources, and storage resources that need to be consumed in the calculation process are also different. Under the condition that the detection reliability and efficiency can be ensured, the on-board processing algorithm needs to be reasonably selected.

The development of intelligent remote sensing satellite originates from the on-orbit processing technology of remote sensing images, which contains a large number of on-board real-time processing algorithms and provides a basic guarantee for the real-time intelligent processing of intelligent remote sensing satellites. With the development of spatial information network and artificial intelligence, the research of intelligent remote sensing satellite has ushered in new opportunities. Spatial information network can provide environmental basis for the operation of intelligent remote sensing satellites and provide basic guarantee for the real-time transmission of intelligent remote sensing satellites. The development of artificial intelligence can greatly improve the intelligence and automation level of on-board data processing and provide strong support for the intelligent development of satellites. Under the background of spatial information network and artificial intelligence, the research on intelligent remote sensing satellites has become an inevitable trend.

8.11 Future Development of a Telegeoprocessing System

Under the traditional application mode of satellite remote sensing system, operation control, reception, processing, and application work independently, and response time is slow. It is difficult to meet the new needs of emergencies and national security with high-quality remote sensing images. For example, the receiving, processing, and distribution of remote sensing images in hours previously should be on-orbit processing and real-time transmission. Instead of GB remote sensing data, now only KB effective remote sensing information is needed for public thin clients such as mobile terminals. With the development of remote sensing technology, people will be able to not only play WeChat but also "play satellites" on mobile phones. That is, commercial remote sensing satellites will provide customized services for ordinary users. In order to achieve this goal, spatial information network and artificial intelligence technologies have provided a powerful boost for satellite remote sensing, and the overall development trend of satellite remote sensing system will also produce fundamental changes because of these two technologies. Figure 8.1 shows the logical relationship between PNTRC (Position, Navigation, Timing,

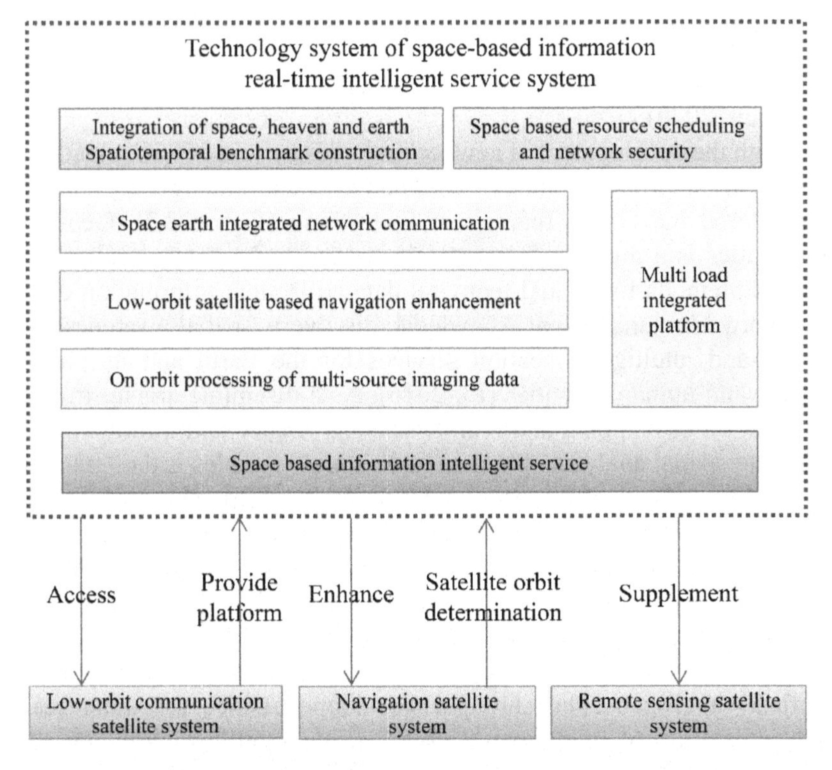

Figure 8.1. Logical relationship between PNTRC and other space-based systems.

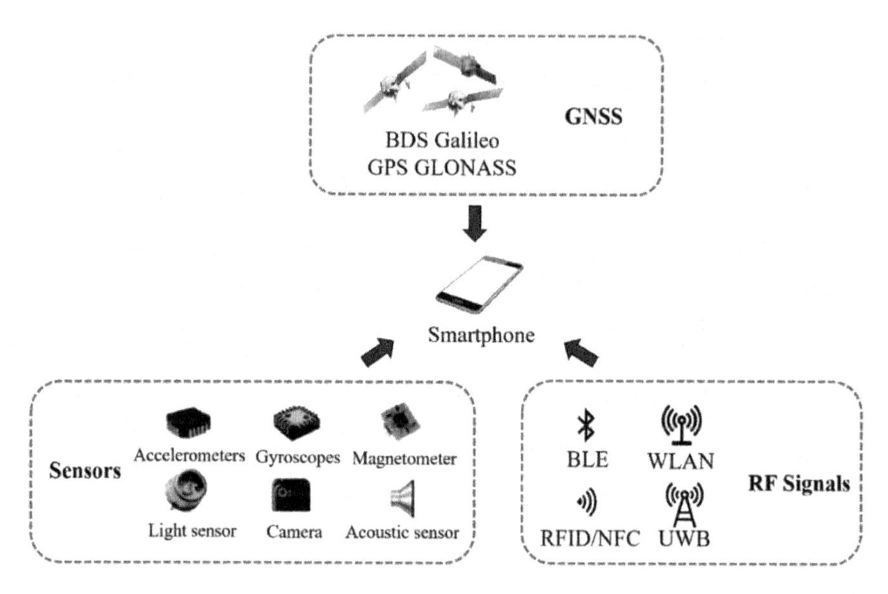

Figure 8.2. Built-in sensors and RF in smartphones.

Remote Sensing, and Communication) and other space-based systems (Li and Shen, 2020).

With the development of new technologies such as 5G/6G, cloud computing, the Internet of Things (IoT), and artificial intelligence, humanity has entered the era of Internet of Everything (Figure 8.2). Geospatial informatics is a multidisciplinary science and technology that integrates various methods for spatial-temporal data collection, information extraction, network management, knowledge discovery, spatial awareness cognition, and intelligent location services for the Earth and its physical objects and human activities (Li, 2016). As a discipline arising from the intersection of mapping and remote sensing science and information science, the spatial analysis of human and natural variables is the foundation of the discipline, and the spatial analysis of geometric and physical variables of the Earth's space can provide the discipline with diverse data and information processing methods. Remote sensing (RS), geographic information system (GIS), and global navigation satellite system (GNSS) are the common tools for measurement, observation, and analysis.

In recent years, the development of spatial information network and artificial intelligence has brought new opportunities for the research and application of intelligent remote sensing satellites. With regard to spatial information network, countries around the world have conducted in-depth research. Typical projects include the United States Transitional Communication Satellite System Program, the SeeMe Program, and the European Space-based Internet Program. Among them, the United States Defense Advance Research Program (DARPA) proposed the SeeMe program integrating communication and remote sensing, which can use the ground mobile terminal to direct the payload operation and receive images, indicating a direction for the development of real-time remote sensing constellation in the future.

The design of intelligent remote sensing satellite adopts the method of software definition, which realizes the functions realized by the traditional sub-system hardware in the form of software, supports the updating of algorithms and parameters, supports the dynamic expansion of functions, and improves the automation and intelligent level of the satellite. Intelligent remote sensing satellites can realize the integrated operation of communication satellites, navigation satellites, and intelligent remote sensing satellites under the environment of spatial information network, which makes satellites connect with each other, expands the scope of application, and solves the problem of single function and isolation between satellites of traditional satellites. The ground only needs to upload the mission requirements to the satellite. The on-board intelligent planning system can solve the optimal planning scheme and obtain the data by combining the satellite state, transit information, and cloud information in time. This method shortens the processing time and obtains the optimal observation scheme combined with the actual situation to meet the needs of emergency situations. The application mode of intelligent remote sensing satellite combines task-driven and event perception. The goal is much clearer, and the application is more intelligent. It solves the problems of unreasonable allocation of satellite and ground resources and low efficiency of emergency response caused by product-driven traditional satellites. The management mode of intelligent remote sensing satellite adopts the form of integration of measurement, control, and remote control services, which improves the problems of processing difficulties and timeliness reduction caused by the separation of measurement, control, and remote control of traditional satellites and lays a foundation for the intelligent and popular development of remote sensing satellites. The intelligent remote sensing satellite can make the intelligent level of the satellite group change from single satellite intelligence to multisatellite cooperative group intelligence. Through the real-time transmission function of the satellite-ground satellite link, the communication, navigation, and remote sensing satellites are connected to realize the development of multisatellite cooperative group intelligence.

Li *et al.* (2022) proposed the concept of Earth Observation Brain (EOB), which combines remote sensing satellites, navigation satellites, and communication satellites to form a multilayer satellite network structure to serve users. In the context of artificial intelligence, various information systems can be organized and applied in the manner of a human brain. The EOB, as an intelligent EO system capable of simulating the cognitive process of the human brain, can automate the following three processes (Li, 2018): (i) acquisition, organization, and storage of massive spatial data; (ii) intelligent spatial data processing, information extraction, and knowledge discovery; and (iii) spatial data-driven applications.

The emergence of EOB, smart city brain, and smart terminal (mobile phone) brain is the inevitable result of the integration and fusion of artificial intelligence, brain cognition, and EO technology in the era of big data, which can promote the intelligent development and application of geospatial information. The concept of mobile surveying continues to evolve and is moving away from mobile mapping and geographic information acquisition applications. The need for new application models (collaborative sensing) and new industry forms (smart connectivity) has led to more opportunities for development in the big data era of building a smart planet and smart cities.

The Digital Twin digitizes people, objects, and events in the city through IoT, GIS, 5G/6G, Building Information Modeling (BIM), and other technologies to build a "virtual city" in cyberspace. Through cloud computing, artificial intelligence, and big data analysis technologies, it can provide situational insights and scientific decisions on urban development and reverse the information obtained to the real city to realize "controlling the real with the virtual", thus forming an urban development pattern with deep learning capabilities, iterative evolution capabilities, and a blend of the real and the virtual.

The intelligent terminal brain can integrate and analyze the distribution, trajectory, and trend of individuals such as "people, vehicles, and things" at the microscopic scale. It can study human behavior and psychology and develop intelligent driving brains and various measurement robots. At the medium scale, we will build digital twins, smart cities, smart traffic, smart public security, smart health, "one map" and "one platform", and other smart social brains, which can analyze and summarize the trajectories of architectural space, urban, and social development and assist in urban planning, resource scheduling, emergency management, etc. (Zhang *et al.*, 2021). At the regional scale, the establishment of national sky-to-earth observation brains can integrate multisource and multiangle information from air, sky, earth, and sea and realize national, global, and even space connectivity. At the macro scale, it can help study the relationship between humans and nature and analyze the distribution, causes, and solutions of population, resources, environment, and disasters. Through the establishment of a multiscale artificial intelligence system, the understanding of the relationship between humans and nature will be taken to a new level and new solutions to many of today's global problems will be available.

At present, remote sensing satellites can provide data support and service help in meteorology, environmental monitoring, resource management, and other professional fields, but the different data and service systems lead to the problem of the joint application in navigation, communication, remote sensing, and other service systems. With the increase of remote sensing data and the improvement of image resolution, people gradually adopt artificial intelligence technology to process remote sensing images, in order to improve the automatic processing ability of images and solve the phenomenon of "isospectral" and "isospectral foreign body" caused by the improvement of resolution. From the early emergence of artificial neural network, support vector machine, to the rapid development of deep learning in recent years, artificial intelligence technology is more and more widely used in the field of remote sensing. It has been applied to target detection, scene classification, semantic segmentation, change detection, and other aspects. The typical algorithms include convolutional neural network and sparse autoencoder, which are mainly used to deal with high resolution and hyperspectral remote sensing images. Companies such as Orbital Insight and Digital Globe in the United States use artificial intelligence, machine learning, and other technologies to automate remote sensing image processing and depth analysis, so as to provide services for defense, security, environment, and other fields. Domestic Aerospace Star Map Co., Ltd. uses machine learning technology to intelligently process remote sensing images and plans to transform remote sensing data information into useful knowledge for social operation by integrating user knowledge. The research center of remote sensing artificial intelligence application technology promotes the intelligent development of remote sensing data by combining artificial intelligence, big data, and other emerging technologies.

It is one of the challenges facing geospatial informatics in the era of Internet of Everything to return to the origin of remote sensing images, resolve semantic ambiguity through image description, take into account the differences between natural and artificial features, and achieve the correct classification of surface targets. Image interpretation, as a basic application of remote sensing data, is a hot topic of research in the field of remote sensing and mapping. For a long time, remote sensing image interpretation is mainly to achieve classification and symbolization of maps, but the multisense nature of classification makes it difficult to produce consistent and correct interpretation results.

At the level of classification systems, the massive amount of multi-source data provided by integrated air-space-sky sensing networks has highlighted the polysemy of natural features while enhancing the dimensionality of surface observations. From these data, researchers in various fields can extract information according to their specialized needs, construct internally semantically consistent classification systems, and build databases corresponding to them. However, different starting points for scientific questions in different fields and subjective cognitive differences between researchers lead to differences in the description of geographical objects in each classification system. The sample sets thus established differ greatly in sample category definition (naming, semantics), hierarchy and compatibility, etc. The openness and scalability of the sample sets are insufficient to support data sharing and comprehensive use among multiple sample sets (Gong *et al.*, 2021), making the results of feature interpretation affected. In the case of land, for example, an agronomist classifies land as hillside grassland based on ground cover, while a forestry expert may plan it as forestable land based on topography, landform, and other features, thus causing the problem of inconsistency between the planned use of the land and its actual use.

In the classification process, neural networks and deep learning have improved the efficiency and generality of the interpretation, but algorithms that can take into account the interpretation of natural and artificial features are still to be developed. In remote sensing interpretation, features are mainly divided into natural features and artificial features. Artificial features are geometric targets with deterministic shapes (e.g., regular-shaped farmland, standard length runways, and planned landscaping), which can be interpreted deterministically through target extraction and change detection algorithms (Gong and Ji, 2018), whereas natural features are generally fractal targets with uncertainties (e.g., natural vegetation distribution, terrain between two and three dimensions, and coastlines whose morphology conforms to the fractal theory (Mandelbrot, 1967)), which are the most important features in remote sensing. Existing interpretation methods are generally oriented towards either natural features or artificial features but not both. Therefore, it is a challenge to consider the differences between natural and artificial features while ensuring accuracy.

To break through the technical bottleneck of remote sensing image interpretation, we need to integrate the existing classification rules, from the classification system based on different rules to the description of remote sensing shared ontology database to meet various needs, and to complete the leap from the map with symbolic classification as the main body to the new product of semantic description under the real 3D real scene plus ontology database. The labels of both natural and man-made features may vary according to the needs of each field and the perceptions of researchers, but their semantics do not change depending on the classification system. Building a database based on ontology semantics from multisource data can realize the description of geographical objects from the mechanism level and establish a shared ontology database that can be applied in various fields. Among them, the map composed of symbols will gradually use VR/AR/IR technology for semantic description to realize the visualization of real 3D landscapes.

The semantic construction based on semantic ontology is carried out by using ontology database and semantic description method so that the interpretation of remote sensing image gradually changes from 2D to 3D real scene and change detection. For the existing feature images and 3D real-world maps, we combine the idea of fractal science, fuse multiple sources of data and construct data cubes to realize the interpretation of natural and artificial targets, realize the transformation from classification system to semantic ontology, and apply the interpretation results to urban planning, environmental evaluation, and human land relationship. Taking the application results of the new generation 3D geographic information platform in the new basic mapping as an example, in the same scene, the integrated data storage and organization of 4D + data, real-time panoramic video, 3D real scene measurement data, level of detail (LOD) model, indoor fine model, BIM, geological layer, 3D pipeline, water environment, and dynamic elements can be realized, and the visual display, interpretation, dynamic analysis and application can be realized. This change will facilitate the application of artificial intelligence in geospatial informatics.

References

Gong Jianya, and Ji Shunping. (2018). Photogrammetry and deep learning. *Journal of Geodesy and Geoinformation Science*, **1**(1): 1–15.

Gong Jianya, Xu Yue, Hu Xiangyun, Jiang Liangcun, and Zhang Mi. (2021). Status analysis and research of sample database for intelligent interpretation of remote sensing image. *Acta Geodaetica et Cartographica Sinica*, **50**(8): 1013–1022.

Li, D. and Shen, X. (2020). Research on the development strategy of real-time and intelligent space-based information service system in china. *Chinese Journal of Engineering Science*, **22**(2): 138–143. DOI: 10.15302/J-SSCAE-2020.02.017.

Li Deren. (2018). Brain cognition and spatial cognition: On integration of geospatial big data and artificial intelligence. *Geomatics and Information Science of Wuhan University*, **43**(12): 1761–1767.

Li Deren, Xu Xiaodi, and Shao Zhenfeng. (2022). On geospatial information science in the era of IoE. *Acta Geodaetica et Cartographica Sinica*, **51**(1): 1–8. DOI: 10.11947/j.AGCS.2022.20210564.

Kawarasaki, S., *et al.* (1991). The role of surgeon in treatment of chronic respiratory failure after pulmonary tuberculosis operations. *Kekkaku: Tuberculosis*, **66**(11): 793–798.

Mandelbrot, B. (1967). How long is the Coast of Britain? Statistical self-similarity and fractional dimension. *Science*, **156**(3775): 636–638.

Peakall, Rod, and Steven N. Handel. (1993). Pollinators discriminate among floral heights of a sexually deceptive orchid: Implications for selection. *Evolution; International Journal of Organic Evolution*, **47**(6): 1681–1687.

Ohashi, T. and Erickson, H. P. (1998). Oligomeric structure and tissue distribution of ficolins from mouse, pig and human. *Archives of Biochemistry and Biophysics*, **360**(2): 223–232.

Wang Mi, Yang Fang. (2019). Intelligent remote sensing satellite and remote sensing image real-time service. *Acta Geodaetica et Cartographica Sinica*, **48**(12): 1586–1594.

Zhang Jixian, Li Haitao, Gu Haiyan, Zhang He, Yang Yi, Tan Xiangrui, Li Miao, and Shen Jing. (2021). Study on man-machine collaborative intelligent extraction for natural resource features. *Acta Geodaetica et Cartographica Sinica*, **50**(8): 1023–1032.

Chapter 9

Telegeoprocessing Applications and Further Developments

9.1 TeleGeomatic — An Introduction

In modern geographical applications, communication and real-time aspects are combined with spatial aspects. TeleGeomatic is a child of Geographical Information Systems and telecommunications. TeleGeomatic can be considered as a new discipline combining positioning systems (GPS), GSM, cartography, GIS, real-time exchange of information between different sites, and also real-time spatial decision-making (Tanzi). In TeleGeomatic, management in real time is integrated with spatial information. These are no more dealing with applications where traditional cartography is the crucial point but where real-time management, sometimes of mission critical type, integrates spatial information. TeleGeomatic is a new discipline characterized by

- use of geographical information systems (GIS),
- use of modern techniques of location (such as GPS),
- exchange of information between multiple sites,
- decision-making in real time, and
- remote updating of facilities.

Figure 9.1 presents TeleGeomatic and its relationships with the various techniques and technologies used. TeleGeomatic is characterized by two aspects: Telegeomonitoring and TeleGeoprocessing. TeleGeoprocessing uses GIS and any communication means for sending necessary

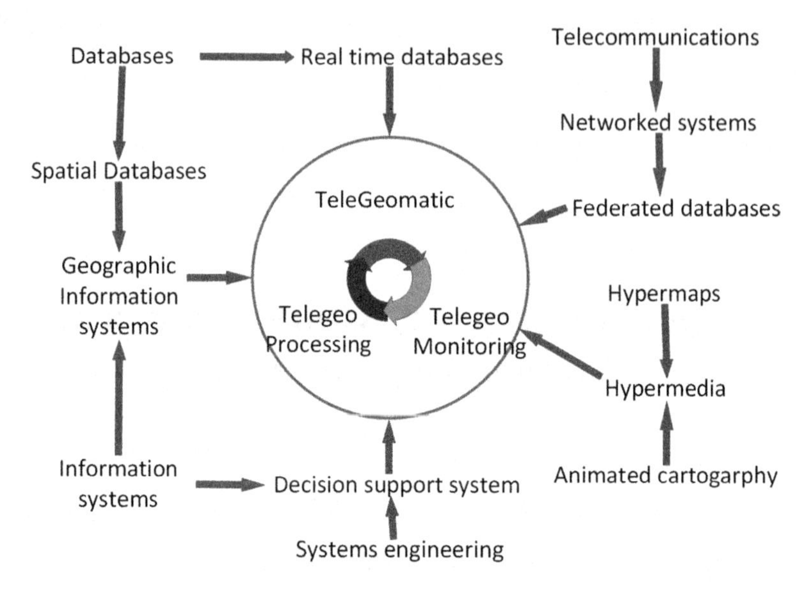

Figure 9.1. TeleGeomatic organization.
Source: Tanzi (2000).

information for analysis. This analysis is done in real time. The final goal of the system is to produce the necessary information for decision in real time. It is necessary to define the notion of real time. In our case, we will use the notion of free real time (FAR Real Time) to define a time of reaction at the human scale. This delay will range between some seconds and some minutes. A TeleGeomonitoring system is composed of a TeleGeoprocessing system, intended for the decision-making. However, it allows execution of actions produced by the decision. For this, it includes a system able to transmit orders and information to the different facilities composing the global system. It must have also the capacity to remote activation, that is to say, of functionality in the classic domain of the automatic device.

TeleGeomatic data are referenced in space and located in time and vary in those dimensions. For example, observation of the evolution of river flood requires spatial and temporal data. To determine the course of the stream, it is necessary to use some spatial information. But the height of water depends on the period of the year and takes the temporal aspect of data into account. Data fluctuate over time. Another characteristic is the duration of the meaning of the data. The life span can be very short or very long and depends on the phenomenon. We must take life span into account for the transmission. The transmission time is never equal to zero. For example, a communication of a signal to any place on the planet Earth using two satellites and one ground control station implies a propagation delay of about one second. In this case, if the used satellites are positioned on a geo-stationary orbit (36,000 km), signals are going to browse about 144,000 kilometers. If the system is not adapted to the life span of data, or if the propagation time is too important, processing will not be able to use this information. Data hold public and private parts. Data security is very important. In risk monitoring, for example, critical data must be protected against unauthorized uses. In certain cases, the use of uncontrolled information can make more damage than the event to which relates data. The protection of people and the respect for the private life and the law impose the protection of data, which are relative to them. Modern technology allows implementing security procedures. But the integration within processings always represents an increase in the complexity of the final application. Data are used as information (data acquisition) or to command or control equipment.

To build a correct representation of TeleGeomatic data, it is necessary to take their characteristics into account:

- *Location in space and time*: Following the previous example, the height of water of the river is varying over time and over space. Every sample corresponds to one very precise moment. Each point that defines the bed of the river must be identified perfectly by its position in the space (x and y, possibly z) and its location in the time which characterizes the precise moment of the measure.
- *Typology and value of data*: The typology and the value of data constitute the kernel of information.
- *Creator and user of the data*: The identification of the data creator gives information on the means of acquisition. The identification of the creator and the user of information's permits to implement security procedures.
- *The quality of data characterizes acquisition means*: The quality of data allows the adjustment of problems or conflict due to the difference in precision of the various means of acquisition. One can mention an inconsistency, for example, between two values measured: one manually and the other using a high precision automatic means.
- *Reliability of data characterizes transmission means*: The reliability of the data characterizes the capacity of transport means used to

accomplish their mission in the prerequisite conditions. This information can also help adjust conflicts or in case of doubtful data values. It allows, with the data quality, to define the confidence that one can grant to information.

The information cycle of the TeleGeomatic data starts at the location of the data acquisition. Let's take an electronic station for traffic data acquisition located on the side of the motorway. An accident occurs on a lane of motorway, and a traffic jam situation is detected.

Conditions of traffic change. The electronic station continues the information acquisition, and after acquisition, data are transmitted to be analyzed into the management center of the motorway. In this example, the analysis is made in the local management center of the motorway, named District. Analysis provides information used in the decision process. For example, a decision might be to reduce the traffic flow. These decisions require actions. To reduce the traffic, it is necessary to turn on the red light that forbids the access to the freeway in accident area. It is also necessary to address information messages to users on the various display panels of the concerned zone. Actions defined by the decision process are transmitted to be executed by the various facilities. On the location of equipments, a control process executes the decided actions. In our example, actions correspond to turning on a red light to stop traffic flow and to display messages. A new cycle starts for estimation of effects of the emergency plan.

Figure 9.2 represents the cycle of information. The first three phases of the cycle (acquisition, analysis, and decision) constitute the system of TeleGeoprocessing. The last two (action and control) constitute the TeleGeomonitoring part of the system. This second can either exist or not exist in a TeleGeomatic system.

Acquired information is usable for various periods along the time. The needs due to the exploitation require uses in few-delayed time. In a complex action plan, the decision can be to require actions, which are not immediate. Actions can be also delayed according to the strategy or to the available resources (i.e., means usable). The same information should be used to constitute the collective memory of the organization. Analysis of this memory information allows the correction of various action plans in order to prevent same kinds of incidents. If that is not possible, we will use this experience to reduce the consequences at the time of the next kind

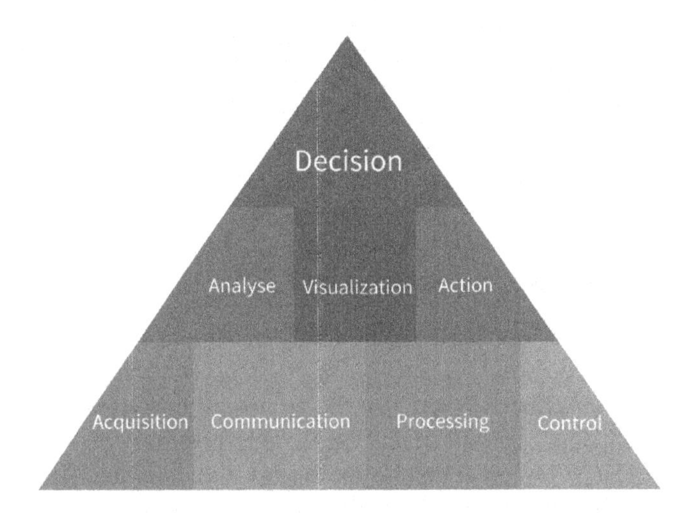

Figure 9.2. A complete cycle of TeleGeomatic information.

of event. The resources definition (i.e., hardware and management) takes incident experiences into account.

Technologic evolutions in several domains, such as electronics and computer engineering, allow the rapid evolution of TeleGeomatic architectures. A few years ago, architectures used were mainly centralized. Users need evolutions. The new technical capacity of the equipment and the computer allows realizing co-operative architectures. The centralized architecture required a control center where data converge to and that communicates with the mobile vehicles and the stationary sensors. In this architecture, only the control unit has a global vision of the system at all moments. The big fragility of this architecture is that if the center of the control does not work anymore, all the complete systems cannot function anymore. In the co-operative architecture, without centralized control, all the components of the system, stationary or mobile, exchange information (between them). They all have a global vision of the context. This architecture is more robust than the centralized architecture. But the quantity of exchanged information is more important. Indeed, one of the priorities is the updating of all data; if the information of a component is different from the one of others, malfunctions occur. Between these two implementations, various architectures are available.

Applications are numerous: traffic monitoring, fleet management, environmental planning, transportation of hazardous materials, or surveillance of technological and natural risks. Functional and architectural similarities exist between all these applications. Computer architectures concerned may be centralized, co-operative, or federated.

9.2 Location-Based Services (LBS) — A Scenario

Location-based services (LBS) have lately become a hot subject. A location-based service (or LBS) in a *cellular telephone network* is a service provided to subscribers based on their current *geographical* locations (http://en.wikipedia.org/wiki/Main_Page). This position can be known by user entry or a *GPS* receiver that he/she carries with him/her. Most often the term implies the use of a *radiolocation* function built into the *cell network* or *handset* that uses *triangulation* between the known *geographic coordinates* of the *base stations* through which the communication takes place.

Location-based services (LBS) provide users of mobile devices with personalized services tailored to their current location. They open a new market for developers, cellular network operators, and service providers to develop and deploy value-added services: advising users of current traffic conditions, supplying routing information, helping them find nearby restaurants, and many more.

Location-based services answer three questions: Where am I? What's around me? How do I get there? They determine the location of the user by using one of several technologies for determining position and then use the location and other information to provide personalized applications and services. As an example, consider a wireless 911 emergency service that determines the caller's location automatically. Such a service would be extremely useful, especially to users who are far from home and don't know local landmarks. Traffic advisories, navigation help including maps and directions, and roadside assistance are natural location-based services. Other services can combine present location with information about personal preferences to help users find food, lodging, and entertainment to fit their tastes and pocketbooks.

There are two basic approaches to implementing location-based services:

- Process location data in a server and deliver results to the device
- Obtain location data for a device-based application that uses it directly.

To discover the location of the device, LBS must use real-time positioning methods. Accuracy depends on the method used. Locations can be expressed in spatial terms or as text descriptions. A spatial location can be expressed in the widely used *latitude-longitude-altitude* coordinate system. Latitude is expressed as 0–90 degrees north or south of the equator and longitude as 0–180 degrees east or west of the prime meridian, which passes through Greenwich, England. Altitude is expressed in meters above sea level. A text description is usually expressed as a street address, including city, postal code, and so on. One implication is that knowledge of the coordinates is owned and controlled by the network operator and not by the end user himself/herself. UK LBS services use a single base station, with a "radius" of inaccuracy, to determine a mobile handset's location. UK networks do not use triangulation.

Applications can call on any of the several types of positioning methods.

- *Using the mobile phone network*: The current cell ID can be used to identify the Base Transceiver Station (BTS) that the device is communicating with and the location of that BTS. Clearly, the accuracy of this method depends on the size of the cell and can be quite inaccurate. A GSM cell may be anywhere from 2 to 20 kilometers in diameter. Other techniques used along with cell ID can achieve accuracy within 150 meters.
- *Using satellites*: The Global Positioning System (GPS), controlled by the US Department of Defense, uses a constellation of 24 satellites orbiting the earth. GPS determines the device's position by calculating differences in the times signals from different satellites take to reach the receiver. GPS signals are encoded, so the mobile device must be equipped with a GPS receiver. GPS is potentially the most accurate method (between 4 and 40 meters if the GPS receiver has a clear view of the sky), but it has some drawbacks: The extra hardware can be costly, consumes battery while in use, and requires some warm-up after a cold start to get an initial fix on visible satellites. It also suffers from "canyon effects" in cities, where satellite visibility is intermittent.
- *Using short-range positioning beacons*: In relatively small areas, such as a single building, a local area network can provide locations along with other services. For example, appropriately equipped devices can use Bluetooth for short-range positioning.

In addition, location methods can connect to a mobile position center that provides an interface to query for the position of the mobile subscriber. The API to the mobile position center is XML-based. While applications can be fully self-contained on the device, it's clear that a wider array of services is possible when a server-side application is part of the overall service. Some applications don't need high accuracy, but others will be useless if the location isn't accurate enough. It's okay for the location of a tourist walking around town to be off by 30 meters, but other applications and services may demand higher accuracy.

9.2.1 *Positioning technology*

9.2.1.1 *Global positioning system (GPS)*

One of the most obvious technologies behind LBS is positioning, with the most widely recognized system being the Global Positioning System. The Global Positioning System, usually called GPS, is the only fully functional satellite navigation system. A constellation of more than two dozen GPS satellites broadcast precise timing signals by radio to GPS receivers, allowing them to accurately determine their location (longitude, latitude, and altitude) in any weather, day or night, anywhere on Earth. GPS has become a vital global utility, indispensable for modern navigation on land, sea, and air around the world, as well as an important tool for map-making and land surveying. GPS also provides an extremely precise time reference, required for telecommunications and some scientific research, including the study of earthquakes.

United States Department of Defense developed the system, officially named NAVSTAR GPS (**Nav**igation **S**ignal **T**iming **a**nd **R**anging **GPS**), and the satellite constellation is managed by the 50th Space Wing at Schriever Air Force Base. Although the cost of maintaining the system is approximately US$400 million per year, including the replacement of aging satellites, GPS is available for free use in civilian applications as a public good. In late 2005, the first in a series of next-generation GPS satellites was added to the constellation, offering several new capabilities, including a second civilian GPS signal called L2C for enhanced accuracy and reliability. In the coming years, additional next-generation satellites will increase coverage of L2C and add a third and fourth civilian signal to the system, as well as advanced military capabilities. The Wide-Area Augmentation System (WAAS), available since August 2000, increases the accuracy of GPS signals to within 2 meters (6 ft) (Brandt, 2002) for compatible receivers. GPS accuracy can be improved further, to about

1 cm (half an inch) over short distances, using techniques such as Differential GPS (DGPS).

The basic principle behind GPS is that a receiver measures the travel time of a pseudo-random code sent from the GPS satellite to the receiver — in practice, around 0.1 s. From this, the receiver can compute the distance (x) to the satellite which places the receiver somewhere on the surface of an imaginary sphere centered on the satellite with radius x. The distance to a second satellite is then measured, narrowing the potential locations of the receiver to an elliptical ring at the intersection of the two spheres. The potential locations of the receiver can be reduced further to just two possible points by incorporating measurements from a third satellite. One of these positions is then disregarded due to it being either too far from Earth's surface or moving at an unrealistic velocity. By taking readings from a fourth satellite, the receiver can be positioned in three dimensions — latitude, longitude, and altitude (D'Roza and Bilchev, 2003).

Although the basic principle is relatively simple, achieving the desired accuracy requires the satellites to maintain precise orbits, which are continually monitored and corrected. Also, each GPS satellite contains four atomic clocks synchronized to a universal time standard. Timing information as well as ephemeris updates — slight corrections to the satellite position table stored in every GPS receiver — are encoded in the satellite's signal. A secondary benefit is that the clock in a GPS receiver operates to atomic clock accuracy but at a fraction of the cost of a real atomic clock.

Differential GPS provides a correction for errors that may have occurred in the satellite signal due to slight delays as the signal passes through the ionosphere and troposphere and multipath errors. Differential GPS comprises a network of land-based reference stations at fixed locations on heavily surveyed sites where a very accurate position can be determined by means of other than GPS. At these reference stations, GPS receivers calculate the position of the site, which is then compared to the actual known position and used to compute an error correction factor for each satellite. Since the GPS satellites are in such a high orbit, any mobile GPS receivers will be using signals that have traveled through virtually the same section of the atmosphere and so contain virtually the same errors as the signals received by the nearest reference station. GPS receivers can therefore apply the same correction factors that were computed by the reference station. Each reference station broadcasts correction factors to differential GPS receivers on a separate radio network. Increasingly, organizations that require this extra accuracy are setting up their own differential reference stations and broadcasting the correction information free for public use.

GPS offers unprecedented positional accuracy but it does have some drawbacks. Since the satellites are in a high orbit and broadcasting over a large area, the signal is very weak. The pseudo-random nature of the signal allows for small footprint antennas, but, because of the weak signal, the receiver needs a reasonably unobstructed view of the sky. This results in GPS receivers being unable to obtain a position fixed inside buildings, under the cover of trees, or even when between tall buildings which restricts the view of the sky — an effect known as the "urban canyon".

9.2.1.2 *Mobile positioning*

There are, however, other means of positioning in addition to GPS. These other technologies are network-based positioning and typically rely on various means of triangulation of the signal from cell sites serving a mobile phone. In addition, the serving cell site can be used as a fix for the location of the user. The terms mobile positioning and mobile location are sometimes used interchangeably in conversation, but they are really two different things. Mobile positioning refers to determining the position of the mobile device. Mobile location refers to the location estimate derived from the mobile positioning operation.

There are various means of mobile positioning, which can be divided into two major categories: network-based and handset-based positioning. The purpose of positioning the mobile is to provide location-based services (LBS), including wireless emergency services.

This category is referred to as "network based" because the mobile network, in conjunction with network-based position determination equipment (PDE), is used to position the mobile device.

One of the easiest means of positioning the mobile user is to leverage the Signalling System #7 (SS7) (http://en.wikipedia.org/wiki/SS7) network to derive location. When a user invokes a service that requires the MSC to launch a message to an LBS residing on an SCP, the MSC may launch an SS7 message containing the cell of origin (COO) or cell ID (of the corresponding cell site currently serving the user). While potentially covering a large area, the COO may be used by LBS to approximate the location of the user. This type of positioning therefore has a large degree of uncertainty that should be taken into account by the LBS application in terms of the required quality of service (QOS). COO is not always available (for example, via SS7 with non-GSM WAP-based services) or does it always meet the QOS requirements of the LBS application. Therefore, network-based (or handset-based) PDE must be employed.

Angle of arrival method involves analysis of the angle of arrival (AOA) of a signal between the mobile phone and the cellular antenna. AOA PDE is used to capture AOA information to make calculations to determine an estimate of the mobile device position. Time of arrival method uses the time of arrival (TOA) of signals between the mobile phone and the cellular antenna. TOA PDE is used to capture time difference of arrival (TDOA) information to make calculations to determine an estimate of the mobile device position. Radio propagation techniques utilize a previously determined mapping of the radio frequency (RF) characteristics to determine an estimate of the mobile device's position. Some hybrid methods of AOA and TOA exist that use the best of both to provide improved positioning.

This category is referred to as "handset based" because the handset itself is the primary means of positioning the user, although the network can be used to provide assistance in acquiring the mobile device and/or making position estimate determinations based on measurement data and handset-based position determination algorithms.

The SIM Toolkit (STK), as an API between the Subscriber Identity Module (SIM) of a GSM mobile phone and an application, provides the means of positioning a mobile unit. Positioning information may be as approximate as COO or more precise through additional means such as the use of the mobile network operation called timing advance (TA) or a procedure called network measurement report (NMR). In all cases, the STK allows for communication between the SIM (which may contain additional algorithms for positioning) and a location server application (which may contain additional algorithms to assist in mobile positioning). STK is a good technique to obtain position information while the mobile device is in the idle state.

Enhanced observed time difference (E-OTD) is what is also referred to as reversed TOA or handset-based TOA. The basic method is employed as with TOA; only the handset is much more actively involved in the positioning process. Specially equipped handsets are required. Perhaps the best known or recognized handset-based PDE is based on the Global Positioning System (GPS). By itself, GPS can be the most accurate (when satellites are acquired/available), but this technology is often enhanced by the network. Assisted GPS (A-GPS) refers to a PDE system that makes use of additional network equipment that is deployed to help acquire the mobile device (much faster than non-assisted GPS) and provide positioning when the A-GPS system is unsuccessful in acquiring any/enough satellites.

9.2.2 *Research in location-based services is still needed*

The rise of location-based services has led to the definition of a new discipline of TeleGeoProcessing, which can be defined as a merging of GIS and telecommunications to support decision-making (Laurini, 2000).

A challenging task in the LBS system engineering is to implement a highly scalable system architecture which can manage moderate-size configurations handling thousands of moving items as well as upper-end configurations handling millions of moving items. A number of research groups have explored different approaches for structuring of location records. Various indexing schemes, such as RT-tree, 3D R-tree, 2+3 R-tree, STR-tree, TB-tree, hashing based index, and TPR-tree, have been proposed (1–6). However, most of the current research activities are centered around single node-based LBS systems. One of the major challenges in the field of LBS system engineering is to achieve a high degree of scalability to handle millions of moving items. Unlike most of the past and current research activities in this area which are pursuing single-node systems, Nah *et al.* (2005) have explored a novel distributed computing architecture devised to handle a very large volume of moving items such as cellular phone users. In order to achieve a reasonable degree of scalability, it is inevitable to exploit distributed and parallel processing based on the use of a cluster of database-centered computing nodes. The architecture named the *Gracefully Aging Location Information System* (GALIS) is a cluster-based distributed computing system architecture that consists of multiple data processors, each dedicated to keeping records relevant to a different geographical zone and a different time zone (Nah *et al.*, 2005).

The architecture named the *Gracefully Aging Location Information System* (GALIS) has been evolving by employing the *Time-triggered Message-triggered Object* (TMO) programming scheme since 2002 (Scheider, 2001; Towles, 2002). This is a cluster-based distributed computing system architecture which consists of multiple data processors, each dedicated to keeping records relevant to a different geographical zone and a different time zone. The architecture is thus capable of dynamic relocation of records among the processors as the items move through different geographical zones and the ages of their past location records keep increasing. In addition, geographical zones assigned to processors can change dynamically for the purpose of load balancing as the population of moving items in a certain region explodes above or shrinks below preset thresholds. Efficient concurrency exploitation is a key factor in making GALIS a very attractive architecture. It is particularly important to facilitate efficient concurrent handling of the never-ending stream of new location-sensing reports of moving items coming from sensors (or relay stations) and the location-based queries coming from various clients. Such data processing frequently involves cooperative actions of distributed database-centered computing nodes and sometimes cooperative reconfiguration of the databases kept in the nodes.

One of the important new challenges in this area is the focus on real-time information. Although companies put different real-time wireless navigation services on the market almost every day, many of these services work inefficiently and are difficult to use: Cognitive aspects of real-time spatial decision-making, people's information needs, and data quality aspects are often neglected. Changes in technology happen so fast that research on critical issues is lagging behind. Research in GI Science, such as human spatial cognition, user interface design, data quality, and spatial analysis, concerning important aspects of TeleGeoProcessing, is needed to facilitate user interaction with real-time navigation services.

The components of a real-time navigation service, i.e., user, client device, server, and database (see Figure 9.3), should fulfill different constraints in order to work efficiently as a whole system. Human spatial

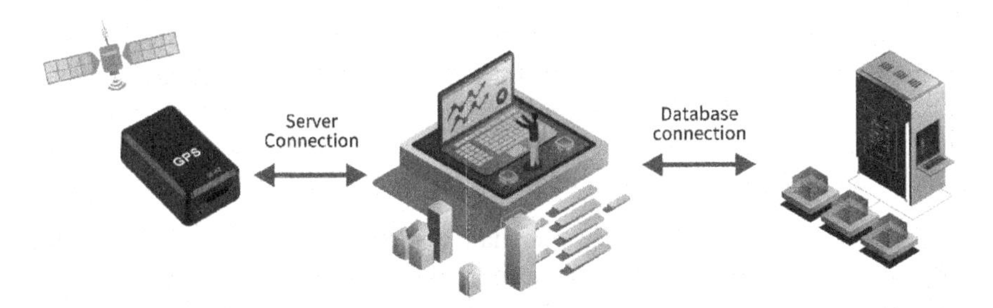

Figure 9.3. Simplified setup of a real-time navigation service (the broken line represents a wireless connection between client device and server).

cognition and way finding are the basis for assigning individual users to user groups. They are both done according to people's personal abilities, tasks, path selection preferences, and information needs. Presentation of information to the user is limited by the capabilities of the client device, i.e., screen size and resolution. Based on rules from generalization and visualization, digital cartography needs to offer alternatives to map modes of communication, e.g., sketches, symbols, or verbal instructions. Issues of data quality are even more complex to handle because contrary to many other GIS applications, real-time navigation services deal with dynamic spatial data, a continuously changing real world, and moving clients. Quality constraints concern both user data (attribute and semantic accuracy of user profile data, resolution of task description, and positional accuracy of user localization) and the underlying cartographic data (completeness and currency, as well as positional and attribute accuracy of the represented objects). Dynamic spatial analysis is needed to develop algorithms for real-time calculation of different paths, including user tracking, path corrections, and error adjustment computations.

9.3 Telegeomonitoring: A Real-Time Safety System

- Traffic management applications
- Fleet management
- Risk assessment on hazardous transport
- Flood monitoring
- Seismic and volcano activities monitoring
- Research and operational uses.

They all need real-time TeleGeomonitoring systems. We cannot present all these applications and we prefer to focus on some of them. More information concerning all these types of applications can be found in Ferguson (2001), Kaplan (2001), Lanier (2001), and Mulligan (2001).

9.3.1 *TeleGeomatic traffic management system*

In traffic management applications, data describe mainly vehicle counting and weather information. Much electronic equipment, distributed along the border of motorway, acquire these data. On classical motorway, several hundreds of acquisition stations can be used. Data collection and computing are done in local technical area. The management center receives information for analysis, makes decision, and transmits data to run the process located on the equipment, corresponding to actions required by the management plan. Principal aspects of this kind of application are real time and availability to support important quantity of information.

In the development of projects phase for the exploitation of freeways, it is necessary to take the traffic capacity into account. To the largest sense of the word, the road capacity is the measure of the efficiency with which the traffic is possible. The use of this notion requires a general knowledge of traffic evolutions and a particular knowledge of flows that can be achieved on the different types of roads at a time and in conditions of exploitation specify (Ditlea, 2001).

It is essential to use a rational method to determine road capacities if one wants to get a good economic and functional utilization of the road network. To the narrow sense of the word, the capacity is here defined as the maximal number of vehicles that can flow out by unit of time on a definite road section with features data. Problems of security are mainly bound to the following factors: the curliness, the declivity, and the width of the emergency strip. They are accentuated strongly when levels of traffic come closer to the capacity of the road. When the flow passes 90% of the theoretical capacity, different studies showed that the accident rate progresses strongly, due to phenomena such as reduction of delay between vehicles and strong growth of the dangerous interval proportion (less than one second). Considering the traffic heterogeneity and the real value scattering around the average, it is necessary to consider that the outflow of the traffic risks to be disrupted from a threshold appraised to 90% of the theoretical. We are now going to see how it is possible to calculate in real time and continuous by the capacity of a road infrastructure. Once this theoretical computation is achieved, it is necessary to compare it with the observed traffic flow. The result of this comparison permits to define actions necessary to assure the safety.

The capacity is defined as the maximal number of vehicles having luck reasonable to be run off during an interval of reference time, on a section of freeway, in meteorological context and feature of traffic given. It can be appraised from the relation's *concentration flow-speed* or of methods operating from the interval between vehicles. The capacity can vary strongly according to the kind of displacement or more precisely by the degree of driver's itinerary knowledge. Abundant research has been achieved in this domain, in particular in the United States: the famous "Highway Capacity Manual", published in 1965 and actualized in 1985 and in 1994 (Ditlea, 2001), makes authority in this domain.

A part of this information is elaborated in the definition phases. They are stocked in databases used by the geographical information system (GIS). Others can be produced by the electronic facilities. The aim of this computation is to establish a representative picture of traffic conditions of a road section. Operator uses two important pieces of information: the maximal theoretical capacity and the position of the traffic in relation to the destabilization threshold.

The operator can use the different measures of traffic engineering to adjust the flow in order to not overpass the destabilization thresholds.

When an incident occurs, real-time information allows verifying how appropriate resources are for the problem. Real-time data simulation tools give a projection of the incident evolution and allow identifying the importance of the incident. Action plan implementation and strategy definition are also based on real-time data. Figure 9.4 shows the real-time management of events. Taking information into account allows to build the incident memory and to enrich the organization experience. In the future, using this experience to modify generic intervention plans will improve the operator's ability to face traffic conditions' modifications. The decision process is composed of steps which are not always sequential.

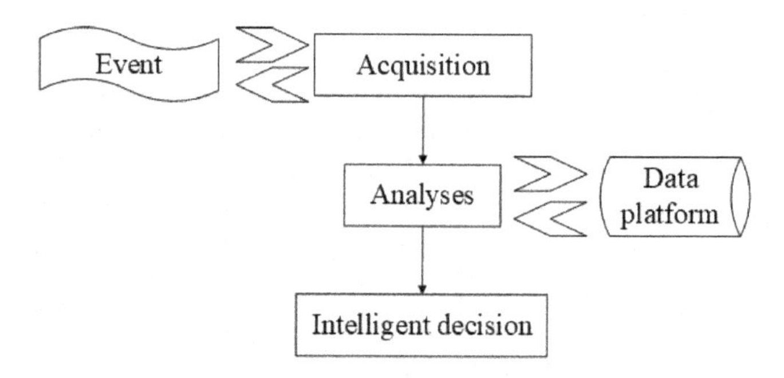

Figure 9.4. Real-time management of events.

Sometimes, backtracking is necessary. The first step is acquisition of the relevant data. The aim is to collect as much as possible information concerning the occurred problematical event.

Next, the conception phase consists in developing one or more *a priori* possible solutions and scenarios. At this step, additional information can be required which implies a second acquisition process (backtracking). The third step consists in choosing the best solution among the list previously built. The final step, the evaluation process, consists in evaluating the previous choices and decisions. The aim of the final step is to correct if necessary the entire reasoning. The set of information is issued from acquisition stations and meteorological stations. Another set of information, for example, concerning road works, is acquired by communication with the existing information system.

Figure 9.5 gives an example of a potential global system architecture for a motorway. This architecture is required to use information issued from traffic data at various levels of the hierarchical motorway organization (Technical Office, District, and Control Station). Information flow and processing according to offices are described in Figure 9.6. The basis office concerns Acquisition Stations in charge of data *acquisition*. Then, Technical Offices have to *concentrate* these data. A *local analysis* is realized by district. Finally, *the Control Station makes data consolidation*.

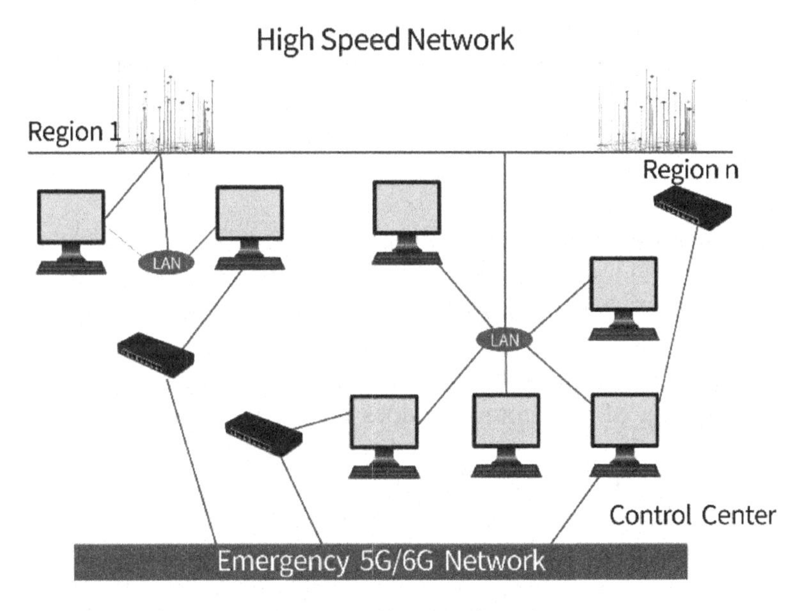

Figure 9.5. System architecture along the motorway.

Data acquisition is realized in real time by sensors (as previously described in the first part) and characterizes each vehicle passing. Acquisition Station is in charge of real-time acquisition and then data

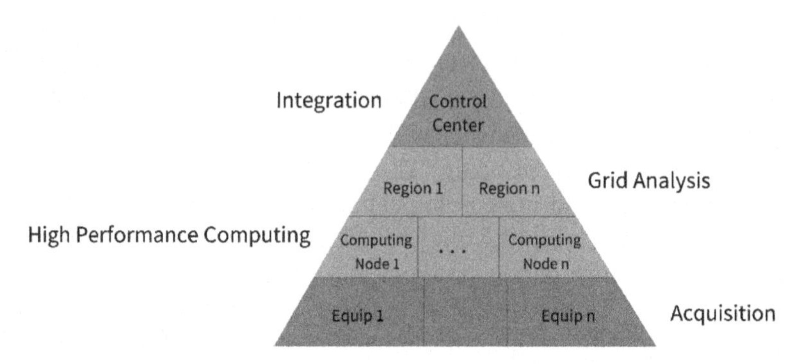

Figure 9.6. Information flow of motorway organization.

transmission to the Technical Offices they depend on. Concentration of acquired data is realized by various Technical Offices. Each station corresponds to a specific geographical area of the motorway. First, data are stored locally and sent every night to the control station by batch processing. Information describing the motorway lanes is stocked in the database. It is necessary to convert the geographical position of the location that one wants to analyze that is expressed in Kilometers Point (PK) in order to make it compatible with the system of co-ordinates of the GIS. Then, the data needed for capacity computation are extracted from the GIS database.

Data local analysis is made by the Districts. A District is an administrative entity representing about 40–50 kilometers of motorway. Every District receives data synthesis from Technical Offices. A cyclical observation of the traffic is made for every station. The frequency of data scanning depends on the alert level of every station. Data consolidation: The Control Station of the motorway does data consolidation. Information concerning road works on the motorway is taken into account by the Control Station. In this aim, the system uses data issued from various external databases belonging to different departments of the motorway like the road works planning department. Taking into account road works information, the Control Station is able to balance synthesis elaborated by districts. The Control Station has a unique and global vision of the entire network of the motorway at one's disposal.

Risk rate estimation of infrastructure allows to build a strategy in order to reduce incident consequences or, if possible, to prevent it. Specific positions must be detected such as positions where the traffic is quite saturated with high-speed vehicles and too low distance between vehicles to avoid pile-up. When such a position is detected, the aim of preventive action is to make drivers reduce their speed. This can be done using along motorways electronic boards displaying dynamic messages. It is also possible to prevent traffic congestion by lane closing using an automatic folding arm (see Bradt, 2002) as existing on Escota's network motorways.

More and more people run and take motorways. Risk of traffic jam and accident increases, and when an accident occurs, it is very difficult to obtain information and to intervene rapidly. The aim of our system is to help real-time motorway managers to prevent accident, and if it is too late, to help them to build their intervention strategy and to follow their implementation.

The aim of the architecture of the real-time information system presented was to optimize the network capacity. Only useful data are sent on the network concerning raw data or data synthesis. Only fit data are given to operators if a pre-alert or alert situation is detected. Data transfer depends on motorway situation (normal: no transmission; pre-alert and alert: transmission) and the system moves automatically from one state to another according to received and calculated data. Data are permanently displayed and, to be more efficient, are graphical. Real-time aspects of the system allow people to react rapidly and sometimes to anticipate a potential crisis.

9.4 Use of Telematics for Emergency and Disaster Management

The general tendency of the annual global economic costs and life losses, related to disasters, rises from year to year. Each year, on average, disasters worldwide kill more than 133,000 people and make more than 140 million homeless. In 1996, financial losses of approx. 440 Billion Dollars were reported by Flannery *et al.* (1997). Progress in disaster prevention, prediction, and warning capabilities allows to decrease the loss of lives and property. However, there is a tendency that disaster costs are still rising.

Nowadays, more information about disasters and hazards is being released than ever before. There have been major improvements by using data from satellites, ground-based networks, and telecommunication.

Moreover, the rapidity of information transmission is increasing due to the Internet. New communication satellites provide opportunities to use e-mail, telephone service, and others in almost any part of the world. The application of remote sensing systems, air-borne and/or from space in combination with GIS systems, allows completely new perspectives for the analysis of disaster cases as well as for training, prevention, and mitigation. In this context, telematics provides the necessary transfer of information and data between different services, sensors, and applications.

Fax and telephone are still the preferred tools of communication between the different disaster-relevant authorities. Daily, thousands of pages are transmitted between administrations via fax. To our knowledge, in rare cases, e-mail, normally without encryption, is used for the regular exchange of information between administrations in national and international organizations. In critical situations and acute disaster cases, modern technologies do not seem to be reliable enough so that a "fall-back" to past technologies is probable.

Generally, an enormous lack of interoperability and interconnectivity in the information and communication flow of disaster management can be found. Besides the communication problem, major gaps are in the field of gathering information, database access, and specialized disaster meta-information availability. Last but not least, we face major problems when accessing and exchanging information in different languages.

Many organizations have the objective to improve disaster information for the purpose of reducing the loss of lives and property. One of the most recent initiatives, the Global Disaster Information Network, GDIN, is promoting the development of new and reliable tools and methods for potential users of natural disaster information sources. GDIN is an initiative of the US State Department; an international conference has recently been held in Mexico City, where a number of international work groups have been established to harmonize global activities in fields such as Technology and Systems Engineering (Kaplan, 2001).

Another important progress can be seen in the establishment of the "Tampere Convention 1998" which provides the legal framework for the use of telecommunications in international humanitarian assistance. Until May 1999, the Convention has been signed by 37 countries (Zimmermann, 1999).

Presently, a multitude of sensors and devices have been developed to collect environmental information and watch negative impacts on the environment; and we hold a large number of databases describing all kinds of negative effects on human health. Although we have an increasing number of tools to treat and analyze data, to forecast impacts, to communicate, and to disseminate information, there are great difficulties to access and use the data since

- good information is expensive,
- datasets are often confidential,
- interoperability is lacking since standards for integration, fusion, and overlaying in order to generate added value are missing, and
- terminology and other languages applied lead to major problems in communicating in difficult and stressing situations.

The whole disaster management cycle requires support from telecommunication. It is evident that more lives and property could be saved if more efforts would be invested in prevention, preparedness, training, and crisis management. Therefore, important efforts have been made to strengthen the pre-disaster stages (i.e., risk analysis tools and early warning of citizens in the disaster-prone area).

They need mainly during all stages of the disaster management communication and information resources (access, fusion, overlay, view, exchange, share, notify, etc.) in an easy to use, fast, reliable, confidential, and cost-effective way.

In order to make the appropriate decisions in this field, the new arising telematics applications are perfectly predestined. Disaster management and information depend to a great extent on the availability and efficiency of telematics (Hadrbolec and Zirm, 1999). The revolutionary development in the field of Telematics during the past years has fostered systems and services in the area of applied disaster information and management. Multimedia applications (text, images, audio, and video) in combination with the Internet break the barriers between traditional and advanced communication technologies. Furthermore, new ways of communication were established like the Internet, satellite-based, and cellular telephone communication.

To go further into the details of the usage of telematics technology for disaster management, we give a brief introduction to the Global Environmental Disaster Information System (GEDIS). GEDIS, the Global Environment Disaster Information System, is a new integrated system for confidential, reliable emergency communication, and coherent disaster management information support. This system will aim to

Figure 9.7. GEDIS overview.

overcome existing information access and communication deficiencies by using advanced user-friendly, Internet-based, multilingual, and multimedia telematic technologies (including tele/videoconferencing and remote interpretation).

GEDIS is a distributed client-server system (Figure 9.7). The internal and external communication flows are confidential and ensured through the use of the best available and/or most appropriate broadband telecommunication technique (fixed, mobile, and satellite based). The basic system is equipped with security, communication (tele and videoconferencing, data exchange sharing, and notification), and data-access utilities, as well as multilingual support and real-time GIS tools. In addition, the basic GEDIS system will allow confidential and reliable communication with other GEDIS systems via a docking mechanism. This opens the possibility for establishing regional, national, and international disaster management networks.

Since an ultimate, all-including disaster information support and communication system is not feasible, a modular approach will be the GEDIS development paradigm. Additional functionality, needed by a user, is provided by tools tailored to specific user needs, to be encapsulated as plug-ins (in short *application-specific plug-ins*). The well-documented, modular, open architecture of GEDIS allows extensions by compatible plug-ins. This approach allows the interoperable interconnection of existing or available disaster specific information resources such as (legally) fixed and quickly deployable monitoring networks, mobile and remote sensors (including satellite data), as well as relevant simulation tools (forecasting and risk assessment), knowledge bases, and decision support systems. A GEDIS implementation will typically include the basic system and one or several related application-specific plug-ins.

GEDIS supports the entire disaster management cycle from prevention to post crisis phase. The system's well-documented, open, scalable, and adaptable architecture allows to tailor a GEDIS implementation for nearly every requirement of a disaster manager by integrating/plugging-in application-specific tools in the GEDIS basic-core system. GEDIS accepts are input from fixed and mobile sensors for real-time data collection, including acquisition on demand of (near) real-time remote sensing data. GEDIS allows to process, integrate, add value, and visualize (using the advanced, real-time, internal GIS capabilities) system-internal and (predominantly) system-external information resources, be they available yet or forthcoming. These include information inventories and online accessible forecasting, risk assessment, and decision support tools. Critical-phase disaster management support is provided by real-time command and control tools. The system employs standard, state-of-the-art (fixed, mobile, and satellite-based) telecommunication networks to ensure a broadband reliable information and communication flow. GEDIS lays the foundation for an Internet-based early warning system for the citizens in a disaster-prone area. GEDIS' open architecture unfolds the market for third-party vendors of relevant tools and datasets and thus contributes to establishing a European standard in the field of disaster management.

Disaster does not respect boundaries of technical disciplines, countries (languages), hierarchies of institutions, etc. Therefore, we experience an urgent need for an integrative unifying approach to serve the very heterogeneous requirements of disaster managers and satisfy the demand for safety and high expectations of the European citizens in a disaster situation. The Tampere Convention will enable an easier international co-operation in actual disaster cases and training. However, there is a lack of systems especially dedicated to the needs of disaster management. GEDIS will hopefully solve some of those problems when it becomes available. One of the potentially driving forces within Europe could be the Commission itself which is preparing the new phase of the IDA 2000 program (Interchange of Data between Administrations, DGIII). In addition, it seems important to aim at an international or even global co-operation which can be intensified by an advanced exchange of information and experience with organizations such as the UN (OCHA and others) and with initiatives such as GDIN from the United States.

9.5 TELEmatics Assisted Handling of FLood Emergencies in Urban Areas (TELEFLEUR)

The objective of the "TELEmatics Assisted Handling of FLood Emergencies in URban Areas (TELEFLEUR)" project is the development of a comprehensive operational system for handling urban flood emergencies that synthesizes cutting edge telematics technology with advanced forecasting of meteorology and hydrology encapsulated in a Decision Support System (DSS) (Koussis and Lagouvardos ,1999).

The consortium that has been formed in the frame of this project is structured as follows:

- National Observatory of Athens (Greece) — NOA
- Universita di Genova, Dipartimento di Fisica (Italy) — DIFI
- International Institute for Infrastructural, Hydraulics and Environmental Engineering (Netherlands) — IHE
- Ingenierbuero Brand Gerdes Sitzmann Wasserwirtschaft GmbH (Germany) — BGS
- Water Corporation Athens, Directorate for Sewage Works (Greece) — WCA
- Regione Liguria — Struttura di Protezione Civile (Italy) — RL
- New Technology System Design Commercial & Construction Co. S.A. (Greece) — G-System.

The utility of a validated, integrated system shall be demonstrated in Athens (Greece) and in Genoa (Italy). Both cities have experienced severe floods in the past due to their complex topography (which plays an important role in the enhancement of precipitation) and to the short hydraulic response time of the basins where the cities are located. Substantial government support has been secured at both sites. Regione Liguria already operates a flood forecasting center in Genoa and is interested in enhancing its capability, in the frame of this project. The Water Company of Athens (WCA), which has the support of the Hellenic Ministry of Environment, Physical Planning, and Public Works, aspires to establish such a capability.

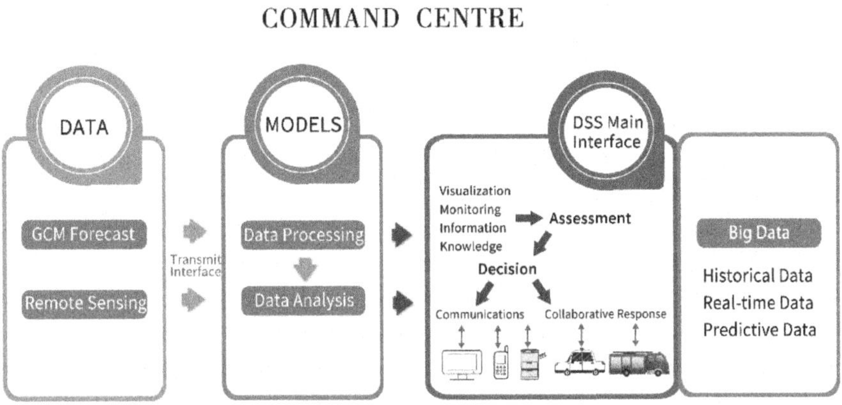

Figure 9.8. Schematic representation of the proposed integrated system.

A schematic representation of the proposed integrated system is given in Figure 9.8. In brief, the system demonstrator will have the capability to

- manage the dynamic information provided by telemetry;
- feed the dynamic data along with the relevant static data (topography, land-use/ground cover, etc.) into models;
- forecast flooding conditions and flood risk estimates;
- assist authorities in decisions regarding emergency measures (DSS), proposing actions for avoiding flood episodes or for minimizing their consequences, and in restoration actions.

The project started in February 1998 and had a duration of 30 months. In order to meet the project objectives, the following steps were followed:

- The user needs for an urban flood forecasting tool have been assessed.
- Up-to-date meteorological models have been tested in Genoa (Italy) and are currently being tested in Athens (Greece).
- The hydrologic/hydraulic analyses of the relevant catchments and associated drainage systems and the validation of the models are in progress.
- The Command Centre already exists in Genoa and has been designed for Athens.

- The communications protocol between the various sources of relevant information (measuring stations, databases, etc.) has been defined for Athens and has been described for Genoa.
- The front end of the Decision Support System of the telematics demonstrator has been developed. For the demonstration phase in Athens, a mock-up of the system will be installed and operated at NOA. The design of the *Command Centre* for the Athens Demonstrator was derived through:
- Interaction with the public authorities responsible for flood emergency handling, especially the Water Company of Athens, partner in this project, and the General Secretariat for Civil Protection (Ministry of Interior).
- Audit of the sources of meteorological and hydrological data and of the means for remote access to and communications with databases and monitoring platforms.
- Audit of the organizational infrastructure required for the co-ordination of the human and technological resources in order to deal with flood emergencies.
- Audit of the flood forecasting system applied by Regione Liguria in Italy (CMIRL, Centro Meteo-Idrologico della Regione Liguria).
- Own analysis of past problematic experiences in Athens.
- Study of the literature and collaboration within the TELEFLEUR Consortium. The main functions of the Command Centre, include the following:
  - receiving meteorological and hydrological data, via telemetry, from all types of sensors (e.g., satellite, weather radar, rain gauges, and stage gauges) and processing these data dynamically;
  - receiving numerical model data;
  - handling of data (input to models/output from models, storage in databases, statistical processing and display, and accessing information from other relevant databases);
  - receiving and sending information about the status of operations and issuing of orders for actions so that communication with authorities and public service organizations can be optimized, throughout the life cycle of an emergency;
  - issuing communications to the individuals, organizations, mass media, and public authorities, each cleared for access at different levels of information, via appropriate, officially approved channels, e.g., the Internet, before, during, and after a severe event;

- forecasting weather and flow on the basis of numerical simulations with meteorological and hydrologic/hydraulic models operating on suitable computing platforms, with appropriate display and printout of results;
- operating a Decision Support System as objective "advisor": evaluate measurements and simulation results data and decide on the guidance to be given to and measures to be taken by the user of the telematics product, in order to handle a projected flood emergency (assist authorities in their actions until restoration to normality, administrative and technical, based on established watch/alert, warning, and alarm status levels). The current stage of the prototype contains several modules that include graphic user interfaces for data input, information display with background maps, graphic information about flooding, generation of flood statements, and a mail composer to send warning statements.

The cost-effectiveness of the system will be judged by calculating the hardware, software, and installation cost of the demonstrator at a particular site. Then, the demonstrator will run with past flood input data; the degree of prevention will thus be evaluated and hence the savings in damages. It will then be possible to calculate a payback period, clearly showing the cost-effectiveness of the system.

9.6 DEDICS — A TeleGeoProcessing and Intelligent Software Agents System for Natural Hazards Prevention and Fighting

The world of Spatial Decision Support Systems (SDSS) for natural hazards prevention and fighting is rich and varied. Hundreds of software products are already available. They supply the users with a variety of information and services in many sectors of activity related to natural hazards management (database, early detection, monitoring, forecasting, and crisis pre-suppression). Considered separately, these products bring great satisfaction to their users. However, natural hazards management needs in terms of exchange of information and services with others SDSS and Information Systems are growing fast, so the placement in the relationship of these tools seems appropriate to bring solutions as the resolution of problems is always more complex. Nevertheless, allowing co-operation between already existing systems in different contexts is not easy. The developer in charge of this work has to overcome difficulties linked with the nature and the content of different products (data formats, operating systems, and programming languages). Despite obvious progress in the level of programming languages (object paradigm), the main problems to solve are related to the consistency, the communication, and the compatibility between heterogeneous systems. Thus, we propose a communicative and cooperative problem-solving approach, based on Telematic and Telegeoprocessing technologies (ground sensors, remote sensing, Geographical Information System, radio-communication, telecommunication, modeling, and spatial analysis) and on an Artificial Intelligence (AI) model (Intelligent Software Agents (ISA)).

Many studies have more or less demonstrated the usefulness of Telegeoprocessing technologies for natural hazards monitoring and prevention, and many authors and research groups have developed prototypes of operational systems including modules for early detection, monitoring, risk assessment, simulation, advising for pre-suppression planning, and natural hazard suppression decision support.

These advanced prototypes are based on the following technologies (Guarnieri *et al.*, 1999):

- Ground sensors for monitoring and early detection: monitoring and early warning of natural hazards are becoming important tools in minimizing risks due to natural hazards. Various types of sensors and cameras have been investigated on their usefulness for ground-based monitoring and early detection systems, such as weather sensors, flood alarms, infrared (IR) cameras (thermal or night cameras), and video cameras.
- Satellite communications are now available for telemetry and telecontrol application requiring portability and fast deployment of the remote acquisition points. These applications are characterized by an absence of infrastructure for communications in case close-to-target data acquisition is wanted, which is normally the case in an emergency situation.
- Remote sensing, Global Positioning Systems, and Geographical Information Systems are now widely used in the field of natural hazards.
- Modeling, Spatial Analysis, and Decision Support: a number of major initiatives have taken place over the past two decades to provide

computational natural hazard modeling tools that enable to simulate environmental processes (forest fires, flooding, landslides, and avalanches).

9.6.1 *Telegeoprocessing thanks to intelligent software agents*

The concept of Intelligent Software Agent (ISA) has been invented to facilitate the co-operation between software packages. The ISA can be a simple procedure or, more often, a more elaborate entity (several tasks in a single program, several processes in the same computer, or several processes on different computers). This concept is a part of a bigger paradigm: Multi-Agents System (MAS). This paradigm is based on the share and the distribution between several agents of the totality of the knowledge and capacities of the reasoning that has an intelligent system. A MAS is therefore a distributed system composed of several agents.

To make a set of Spatial Decision Support Systems (SDSS) more efficient, it is necessary to conceive and to implement methods and computer models which will assist SDSS to cooperate and to coordinate their actions in the framework of a complex problem-solving process. The efficiency of this co-operation depends on the quality of reasoning and on the communication language that will be used inside the model of an agent. The reasoning will be used inside the ISA to allow them to take appropriate decisions concerning the co-operation. The communication language is used to allow the exchange of information between SDSS in a heterogeneous environment. The heterogeneity of the environment can be considered at several levels: heterogeneity between platforms (PC, MAC, and workstations) and heterogeneity between the database and the representation of these data in each software.

9.6.2 *Demonstration on forest fire prevention and fighting*

In order to demonstrate both the capabilities of Telegeoprocessing and Intelligent Softwares Agents, two demonstrations have been organized in July and September 1998. The test location was in the Sophia Antipolis Science Park, a technological and research park, located in the South of Europe, on the French Riviera, between Nice and Cannes. Sophia Antipolis covers a green space equivalent to a quarter of the surface area

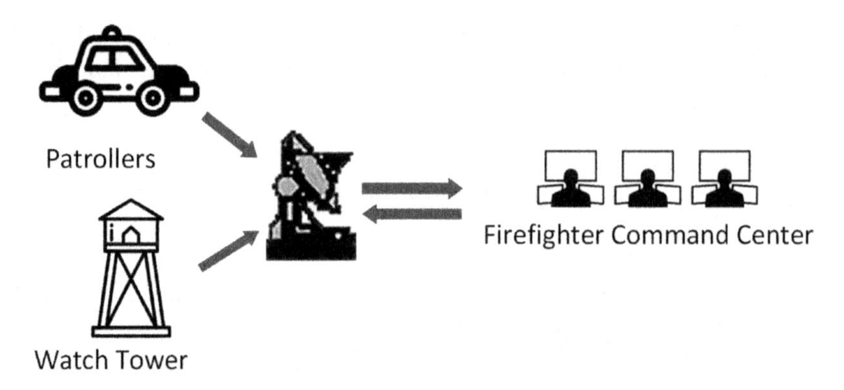

Figure 9.9. The actors involved in the demonstration.

of Paris, currently 5750 acres. These features represent a typical case of interface between natural and urban areas in the Mediterranean region. The presence in the same place of important companies and large forest areas increases the vulnerability and thus the risk of forest fire.

The following three actors are involved in the demonstration (see Figure 9.9):

- The Firefighters Command Center (a Command Center is located in the heart of Sophia Antipolis), where several specialized SDSS are installed.
- The Firefighter Patrols are represented by a truck (first intervention) equipped with a GPS and radio communication system.
- A watchtower where an autonomous early detection system is installed and which communicates with the Firefighters Command Center through a satellite link.

Five applications specialized in forest fire decision support are considered. The communication between the five Decision Support Systems is done thanks to satellite, telephone, radio, and internet/ethernet networks. We give a brief description of these systems:

- **WILFRIED**: The Wilfried system contains four dedicated modules using the data available in the firefighter's GIS: Map Viewer, Climatic Risk Assessment, Land Use Risk Assessment, and Preferential Spread Area Modules.

- **LogBook SYSTEM (LB/MCI):** This system allows the user to store all the messages received (in the Firefighter Command Center) and forwarded (to firefighters on the field) during a forest fire. These messages are organized in a Database Management System thanks to a data-conceptual model. This system provides also a synthesis of each event, is automatically updated with each new message, and proposes some functions to contribute to the backtracking process.
- **METEOROLOGICAL MONITORING SYSTEM**: This module is based on a real-time data acquisition system. The system is connected (via modems and a telephone line) to a network of 21 weather sensors throughout the Alpes-Maritimes region.
- **FLORINUS**: This is a system specialized in communication and exchange of information between the different firemen groups on the field (Patrollers). It is used by the different kinds of operational actors to give them an exact image of events and locations (thanks to a GPS) and to provide information when and where it is needed.
- **AFFIRM**: This is a prototype for early and reliable wildfire detection. The system consists of a network of autonomous terminals disseminated in the forest area and linked to the Firefighter Command Center through a satellite communication link.

This demonstration has given the opportunity to present the possibilities of Telegeoprocessing and Telematic for supporting the management of emergencies and more precisely forest fires. The introduction of prototypes of DSS components in the operational environment has been achieved and the interaction with operators has permitted to improve the performance of these applications and to demonstrate their ability to support emergency management tasks. During the discussion at the end of the demonstration, the general impression was positive and it can be concluded that this project is successful, but it should be seen as the first phase of implementation. Several attendees have proposed that a second phase should start from these working prototypes, to develop a real operational system, integrated with emergency management services.

This chapter has briefly described how both Telegeoprocessing and Intelligent Software concepts and technologies can improve decision-making in natural hazards prevention and fighting. Nevertheless, this paper indicates only some main problems and criteria for the development of cooperative Decision Support Systems. A pilot application has validated the methodological and technological approach. Another pilot application, dedicated to forest fire too, is in test since July 1998 in Crete (Greece) in collaboration with the local forest fire services. As far as further development is concerned, it has been planned to

- implement these technologies in the field of seismic risk;
- concentrate our effort on a more advanced prototype of Intelligent Software Agent able to manage in the same time more than one DSS and have high reasoning capabilities;
- enter, as soon as possiblc, the project in the industrialization phase.

9.7 Telematics Applications of the Danube Accident Emergency Warning System

In the 1980s and 1990s, many accidental water pollution events were identified in the Danube River Basin. A significant number of these were international, while still others went unreported. In some countries, these accidents on occasion required the shut down of drinking water intakes or other precautionary measures to be taken. A clear need appeared to improve the early information about such events, especially in the case of their transboundary impacts. Countries that lie in the middle and lower part of the basin are the most at risk in this respect.

The strategic approach to this problem was to establish a system of international co-operation in the Danube River Basin which is able to provide early information on transboundary river pollution incidents (Pintér, 1999). The responsible authorities of the countries affected by the impacts of such pollution events can use this information to assist the effective control and damage prevention activities on a national level. The Rhine International Alert and Warning System which has been in operation for many years as well as the recently developed system for the Elbe River clearly demonstrated the need for accidental pollution warning systems in river basins where important water uses are in operation (Rhine, 1987). This is why the establishment of the Danube AEWS was one of the high priority actions of the Environmental Programme for the Danube River Basin (EPDRB) and was supported by the Governments of the Danube countries with the financial support of the European Union's Phare Programme.

The main objective of the Danube AEWS is to increase the safety of population and protect the environment in case of water pollution accidents, which have adverse transboundary impacts on the River Danube or its tributaries. For this purpose, a basin-wide fast information system on emergencies caused by transboundary pollution incidents was established in the Danube Basin. The provision of such information enables the competent authorities to act in time to protect sensitive water users. The DAEWS was planned to cover the whole catchment area of the River Danube, including all of its significant tributaries.

The set up of the DAEWS is in accordance with and builds on the relevant and adopted multilateral conventions and declarations concerning the Danube or transboundary rivers. The Convention on Co-operation for Protection and Sustainable Use of the River Danube (the provisions on the establishment of communication, warning, and alarm systems are stated in Article 16.) is an important one among the related international agreements as well as the conventions agreed under the UN/ECE framework. The DAEWS is also based as much as possible on the bilateral agreements that already exist between various Danube countries.

The set up of the Danube AEWS is based on the principle of one national center (called PIAC, Principal International Alert Centre) for each of the Danube countries, having the same responsibility and functionality (Delft, 1994). There are at present nine fully equipped PIACs (Germany: Passau, Austria: Tulln, Czech Republic: Brno, Slovakia: Bratislava, Hungary: Budapest, Slovenia: Ljubljana, Croatia: Zagreb, Romania: Bucharest, and Bulgaria: Sofia), all in operational stage. Three other PIACs are under establishment in the lower Danube Basin to serve the countries Ukraine and Moldova. The geographical scope of the DAEWS unfortunately has a territorial "gap" at present. Due to the recent political situation, Yugoslavia is not part of the system in the Danube Basin. The design of the system however is such that it can be easily extended in the future, allowing the integration of the remaining areas.

The Accident Emergency Warning System Sub-Group of the Danube countries designed and directed the development activities of the DAEWS (Hartong, 1994), with the essential technical support of EU consultants. The practical experiences of the design and operation of the Rhine system were taken into consideration and utilized. The Danube Program Coordination Unit in Vienna harmonized the international contributions during the implementation period (EPDRB-PCU, 1996.).

Each PIAC in the co-operating countries has three Units to perform the following tasks during emergency periods:

- Communication on a reported sudden pollution of the Danube river basin waters;
- Expert involvement to assess the effects or impact of the reported accidental pollution;
- Decision-making on further actions (local or international warnings) to be taken.

To assist the proper management of the above tasks, the computer network established at PIACs was suited to the systems in use in the different countries (Novell, etc.). Direct access was established to the national water quality and hydrological data banks. International Operations Manual ensures the standardized operation of the System (Manual, 1997). The operation of the PIACs during emergency periods is supported by the following tools:

- Satellite communication system provides safe and independent transmission of standard messages (warnings) between the PIACs on the planned routes;
- Database of dangerous chemicals pollutants;
- Danube Basin Alarm Model (DBAM). The model system provides an important tool for the impact assessment of accidental water pollution incidents (DBAM, 1997).

During the last two years of operation since the inauguration of the DAEWS, only eight accidental water pollution incidents were recorded in the river basin by the system, much less than expected. Five of these incidents were caused by oil pollution of the River Danube, and in other cases, the pollutants were salt, detergents, and pesticide (Pinter, 1999). The System reacted properly during these emergency periods, sending the necessary messages to downstream countries. The basic information about the pollution incidents was provided for the PIACs in each country by the competent local authorities for water pollution control.

The main fields of outputs coming from the development activities of the Danube AEWS, in which the results could be further utilized as "good practice", were identified as follows (ENWAP, 1999):

- Data collection and preprocessing, data management, Output: system for the management of information flow on a national level within the units of the PIAC and on an international level between the different PIACs in the Danube River Basin. The user-friendly software provides options for each country to use its home language while the language of international communication is English.

- Models and decision support systems, Project output: DBAM Danube Basin Alarm Model applied for each Expert Unit of the PIACs as a tool for decision support. The model system covers the whole catchment area of River Danube and its important tributaries providing capabilities to assess water pollution impact characteristics, like time of travel and maximum concentrations along the affected river stretch.
- Telecommunication, Project output: basin-wide satellite communication and alarm system for the immediate receive and transmission of messages on accidental pollution incidents. Pagers connected to the system provide details on emergencies for personnel alerted.

9.8 Tele-immersion — A Positive Alternative of the Future

Teleimmersion is being developed as a prototype application for the new Internet2 research consortium. Beyond improving on videoconferencing, tele-immersion was conceived as an ideal application for driving network-engineering research, specifically for Internet2, the primary research consortium for advanced network studies in the USA. Tele-immersion involves monumental improvement in a host of computing and communication technologies, developments that could eventually lead to a variety of spin-off inventions. Tele-immersive environments will therefore facilitate interaction not only between users themselves but also between users and computer-generated models and simulations. This will require expanding the boundaries of computer vision, tracking, display, and rendering technologies.

Tele-immersion is a new medium of human interaction that creates illusion that the user is in the same physical space like the other participants, although in reality other participants are thousands of miles away. This is a new telecommunication medium that combines aspects of virtual reality with 3D videoconferencing.

Tele-immersion enables users at geographically distributed sites to collaborate in shared, simulated, hybrid environment as if they were in the same physical room.

Tele-immersion is the ultimate synthesis of media technologies:

- 3D environment scanning — 3D reconstruction based on display;
- projective and display technologies;
- tracking technologies;
- audio technologies;
- robotics and haptics;
- and powerful networking.

The considerable requirements for a tele-immersion system, such as high bandwidth, low latency, and low latency variation (jitter), make it one of the most challenging network applications. This application is therefore considered to be an ideal driver for the research age of the Internet2 community (http://www.internet2.edu). If network can support network requirements for tele-immersive sessions, it is very probable that every other future network application will be also supported. So, the program (initiative) related to realization of tele-immersion could be understood like the USA national politic of ensuring adequate standards and suitable infrastructure solutions for the future.

There are two basic points of view regarding tele-immersion. The first approach considers that tele-immersion is a model that is a three-dimensional, avatar-like, graphical representation of the participants prepared in advance. The second approach considers tele-immersion as a representation of virtualized reality based on 3D reconstruction of real people (participants) and objects as it is explained in this article.

The biggest challenge of tele-immersive system was the creation of an adequate way of visual sensing people and their environment. Tele-immersion is going to give all participants during the tele-immersive (TI) session a sense that they are in the same physical room although they are thousands of miles apart from each other. This illusion is achieved by 3D acquisition of the scene performed using the method "sea of cameras" where the angles of a bigger number of cameras are being mutually compared. Every camera inside of the "sea of cameras" captures visual features from its angle. By comparison of angles and using the algorithms, these captured data can be combined as a three-dimensional model of the scene.

Tele-immersion expects that the workspace (telecubicle — the next-generation immersive interface) contains a person in the foreground interacting with remote users and a background scene which will remain more or less constant for the duration of a TI session. The process of 3D acquisition of three-dimensional sculptures of the people (foreground — dynamic part of the scene) has to be very quick because TI communication has to be performed in real time. The "sea of cameras" performs real-time acquisition of the foreground with a frame rate of 1–3 frames per second (fps), resolution of 320 × 240 pixels, and volume of 1 m^3. This frame rate requires the level of processing of about 15–25 Megabits per second. This level of processing is needed only for 3D acquisition of the foreground (TI participants and their movements). The process of 3D acquisition of the user surrounding environment (background — static part of the scene) is being performed before TI session and offline transmitted to another participating site. For 3D acquisition of the background, the TI system uses laser scanning with a frame rate of about 1 frame every 20–30 minutes under high resolution and volume of the room. This is the reason why the background of the scene has higher fidelity to reality and the scene foreground has lower fidelity to reality when they are displayed at a remote participating site. While the local user moves in the foreground during the session (dynamic part of the scene), the TI system needs a method to segment out the static part of the scene.

After separation of 3D acquisition of the scene (background — offline, foreground — in real time), reconstructed 3D data of the foreground are transmitted in real time to another participating TI site where they are combined in one united scene with a stored and reconstructed offline transmitted 3D data of the background. After that, this reconstructed and synthesized 3D scene is being rendered, projected, and displayed on stereo-immersive surfaces using stereoscopical projectors. Frame rate for reconstruction and display has to be around 20–25 frames per second to maintain the quality of illusion. The key feature of visually compelling tele-presence environment is that the scene is displayed stereoscopically and the rendered view changes according to the viewpoint of the user's head. The remote scene is always projected from the viewpoint of the local user as if he is looking through a window into the remote scene. Due to this, the participant has to wear a highball ceiling head tracker for the measurement of user's head position and 3D stereo glasses to view remote stereo-displayed 3D scene.

On the assumption that each TI site is just one person, without surrounding static part of tele-cubicle and with acquisition frame rate of around 2–3 frames per second needed, the TI bandwidth will be about 20 Megabits per second (Mbps) with peaks up to 80 Mbps per TI site.

Besides bandwidth, tele-immersion needs that a network latency is inside a time period of 30–50 milliseconds. Longer delays result in user fatigue and disorientation, a degradation of the illusion, and in the worst case, nausea. This means that tele-immersion needs to be performed over the high-performance networks like Abilene (http://www.abilene.internet2.edu) or vBNS (http://www.vbns.net) network in the USA.

In the background of the scene-capture system, the main components of the TI system are computers, network services, and display and interaction devices. Tele-immersion requires a cluster of eight 2GHz Pentium processors with shared memory for processing one trio of cameras inside the "sea of cameras" in real time. For the minimum of usual conversation between TI participants, tele-immersive system needs at least seven cameras although tele-immersion will need at least 60 cameras for most demanding tele-immersive uses (surgical demonstrations, consultations, and trainings). Tele-immersion also requires processing power for re-synthesis and re-renderings of the scene as the local user's head and viewpoint of the remote scene is moving around (one processor in combination with graphic cards for each eye to maintain the sense of depth for the displayed 3D scene). Extra processors are needed for other tasks, such as combining results from each trio of cameras (5), measuring the user's head movements, maintaining the user interface, functioning of virtual objects simulation, and minimizing network latency.

Tele-immersive environments have potential to significantly change educational, scientific, and manufacturing paradigms. They will show their full strength in the systems where having 3D reconstructed "real" objects coupled with 3D virtual objects is crucial for the successful fulfillment of the tasks.

Tele-immersion will make the following possible for scientists and researchers:

- "Real" co-operation on geographically distributed sites that is going to result in time savings, better communication, and awareness of achieved results;
- To bring together scientific teams which will be more geographically independent because all experiments and empiric research will take place in virtual space between them.

In a tele-immersive environment, computers recognize the presence and movements of individuals and both physical and virtual objects, track those individuals and objects, and project them in realistic, multiple, geographically distributed immersive environments on stereo-immersive surfaces. This requires sampling and re-synthesis of the physical environment as well as the users' faces and bodies, which is a new challenge that will move the range of emerging technologies, such as scene depth extraction and warp rendering, to the next level. This new paradigm for human–computer interaction falls into the category of the most advanced network applications and, as such, it is the ultimate technical challenge for Internet2. Furthermore, a secondary objective is to accelerate the development of better tools for research and educational collaboration and promote advances in virtual environment research, which means that this plan will be woven into the fabric of many of the Next-Generation Internet applications (http://www.ngi.gov). Specific aspects and components of tele-immersion are going to be improved in the future and will enable users to achieve a compelling TI experience. With the development of Internet2 and its high-performance networks (Abilene and vBNS), international co-operation, and the use of supercomputing resources (SCC's), it will be possible to improve the characteristics of tele-immersive systems. While the system characteristics for tele-immersion are going to improve, prices of compounding TI components are going to fall and that is going to open the space for the stronger use of TI systems outside the scientific and research laboratories.

References

Bradt, S. (2002, November). Tele-immersion system is first "network computer" with input, processing and output in different locations. University of Pennsylvania. *Proceeding of the TIEMEC'95*. http://www.upenn.edu/pennnews/releases/2002/Q4/teleimmersion.html. Accessed on 3/15/2003.

Buisson, L. (1995). Decision support systems for the prevention of slope related natural hazards, a personal view of the French situation. In *Proceeding of the TIEMEC'95*, Nice, May 1995.

DBAM. (1997): EU/AR/102A/91 Phare Project of the Applied Research Programme "Development of a Danube Alarm Model". Prepared by a Consortium led by VITUKI Plc., Budapest.

Delft. (1994). Environmental Programme for the Danube River Basin. Accident Emergency Warning System. Set-up of the System. Final Report. Author: Hans Hartong. Delft, The Netherlands.

Ditlea, S. (2001). Tele-immersion: Tomorrow's teleconferencing. *Computer Graphics World*, January 2001. http://www.cs.unc.edu/Research/ootf/stc/office/media/pdf/CGW_January2001.pdf. Accessed on 3/15/2003.

EPDRB-PCU. (1996). Environmental Programme for the Danube River Basin. Annual Report 1996. Danube Programme Coordination Unit, Vienna, August 1997.

Ferguson C. (2001). Van Dam project hopes to marry 3-D graphics with interactive electronic books that one day may train surgeons using virtual reality. Brown University (George Street Journal), October 2001. http://www.brown.edu/Administration/George_Street_Journal/vol26/26GSJ08f.html. Accessed on 3/15/2003.

Flannery Terrance, and Robert Winokur, J. (1997). Harnessing Information and Technology for Disaster Management. Task Force Report.

Guiol, R. (1994). Neutralisation automatique de voies rapides et autoroutes. Revue des Associations des Ingénieurs et Anciens Elèves de l'Ecole Nationalc dcs Ponts et Chaussées, PCM LE PONT n°12, Décembre 1994, pp. 25–27

Kaplan, K. (2001, February). A virtual world is taking shape in research labs. *La Times*. http://imsc.usc.edu/press/pdfs/IMSC_LATimes_Web_2_5_01.pdf. Accessed on 03/15/2003.

Lanier, J. (2001). Virtually there — Three — Dimensional tele-immersion may eventually bring the world to your desk. *Scientific American*, April 2001. http://www.cs.unc.edu/Research/ootf/stc/office/media/pdf/sciam_20010401.pdf. Accessed on 03/15/2003.

Laurini, R. (2000). A short introduction to TeleGeoProcessing and TeleGeoMonitoring. In Laurini, R., and Tanzi, T. (Eds.), *TeleGeo'2000*, Sophia Antipolis, France, pp. 1–12.

Mulligan, J. and Daniilidis, K. (2001). Real time trinocular stereo for teleimmersion. In *International Conference on Image Processing (ICIP'01)*; October 2001. http://www.cis.upenn.edu/~janem/ps/mulligan01icip.pdf. Accessed on 3/15/2003.

Nah, Y., Lee, J., Lee, W. J., Lee, H., Kim, M. H., and Han, K.-J. (2005). Distributed scalable location data management system based on the GALIS architecture. In *Proceedings of the 10th IEEE International Workshop on Object-Oriented Real-Time Dependable Systems (WORDS'05).*

National Research Council, Board on Natural Disasters, Commission on Geosciences, Environment and Resources. (1999). *Reducing Disaster Losses through better Information.* Washington D.C.: National Academy Press.

Schneider, M. (2001). Pittsburgh Supercomputing Center to Partner in Quantum Molecular Dynamics and Tele-Immersion. Pittsburgh Supercomputing Center, October 2001. http://www.psc.edu/publicinfo/news/2001/nsf-grants100101.html. Accessed on 03/15/2003.

Tanzi, T. and Servigne, S. (1997). A Real-time GIS for accident prevention on toll motorways. In *Proceedings of JEC'97, Joint European Conference an*

Exhibition on Geographical Information, Vienna, Austria, April 16–18, 1997, pp. 42–50.

Tanzi, T., Guiol, R., Laurini, R., and Servigne, S. (1997). Risk rate evaluation for motorway management. In Andersen, V., and Hansen, V. (Eds.), *Proceedings of the International Emergency Management Society Conference (TIEMS 1997)*, Copenhagen, June 10–13, 1997, pp. 125–134. *Publié dans Safety Sciences*, **29**: 1–15, 1998.

Tanzi, T. and Servigne, S. (1998). A real time information system for motorway safety management. In *Proceedings of the International Emergency Management Society Conference (TIEMS 1998)*, Washington, D.C.

Tanzi, T., Boulmakoul, A., Laurini, R., and Servigne, S. (1998). Un nouvel indicateur pour le calcul du risque d'une infrastructure autoroutière. In *2nd International Conference on Applied Mathematics and Engineering Sciences, CIMASI'98*. Casablanca. 27–29 Octobre 1998.

Towles, H., Wei-Chao, C., Ruigang, Y., Sang-Uok, K., Fuchs, H., Kelshikar, N., Mulligan, J., Daniilidis, K., Holden, L., Zeleznik, B., Sadagic, A., and Lanier, J. (2002). 3D Tele-Collaboration Over Internet2. In *International Workshop on Immersive Telepresence (ITP2002)*; December 2002. http://www.cs.unc.edu/Research/ootf/publications/Towles_ITP02.pdf. Accessed on 3/15/2003.

Transportation Research Board. (1995). Highway Capacity Manual, Special Report 209. Washington D.C.: US Transportation Research Board.

Zimmermann Hans. (1998). The role of telecommunication and the tampere convention in disaster management. In *2nd GDIN Conference*, Mexico City, May 11–14, 1998.

Zirm Konrad. (1998). The concept of GEDIS, the global environmental disaster information system. In *2nd GDIN Conference*, Mexico City, May 11–14, 1998.

Index

Printed in the United States
by Baker & Taylor Publisher Services